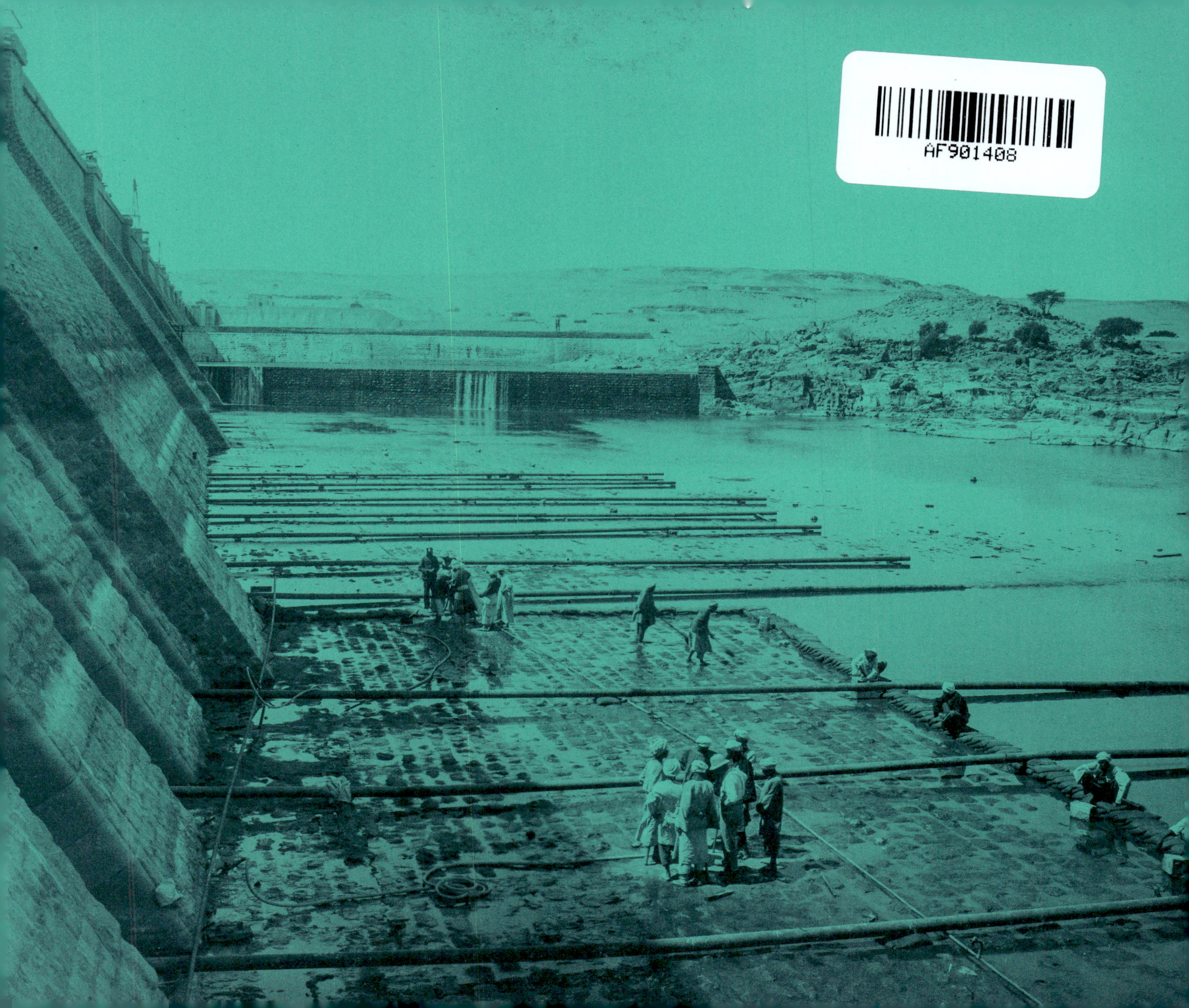

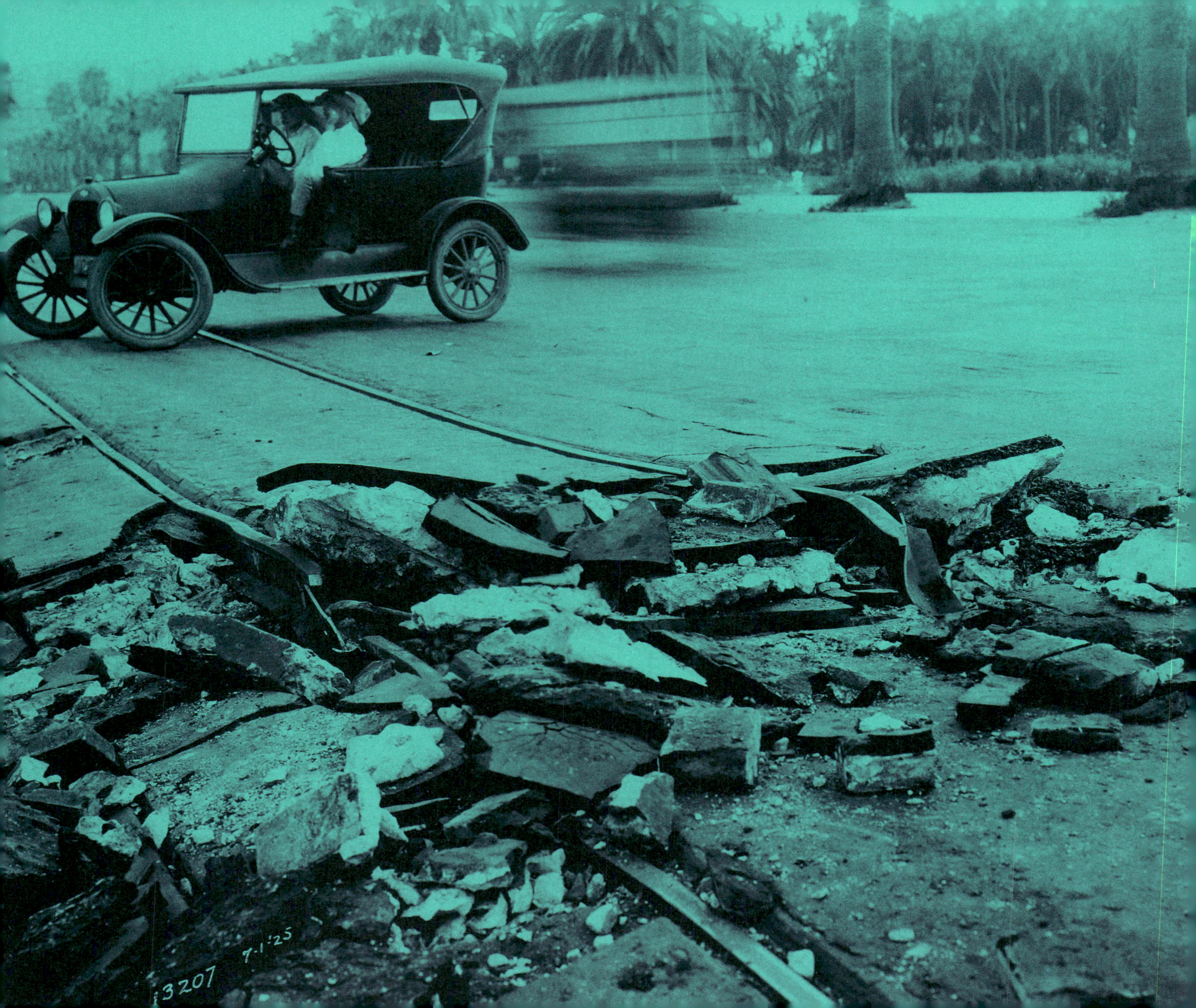
13207
7-1-'25

Worlds Below

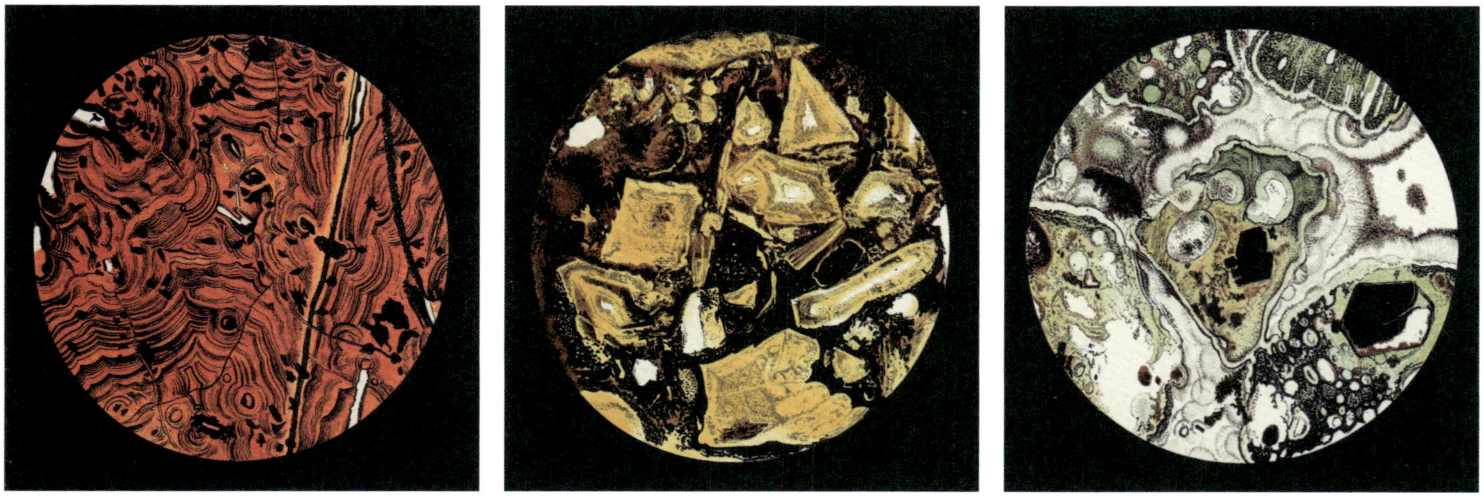

↑ Sections of manganese nodules dredged from the South Pacific in *Report on the Scientific Results of the Voyage of HMS* Challenger *During the Years 1873–1876*, 1882–95. Charles Wyville Thomsen (1830–1882) and John Murray (1841–1914)

Que yo, señora, nací	Because I, my Lady, was born
en la América abundante,	in abundant America,
compatriota del oro,	compatriot of gold,
paisana de los metales,	fellow countrywoman of precious metals,
adonde el común sustento	where common nourishment
se da casi tan de balde,	is given almost for free,
que en ninguna parte màs	because nowhere else
se ostenta la tierra madre.	does Mother Earth flaunt herself more.

—Sor Juana Inés de la Cruz, *Romance 37: Grande Duquesa de Aveyro*, ca. 1680

Invisible fish swim this ghost ocean now described by waves of sand, by water-worn rock. Soon the fish will learn to walk. Then humans will come ashore and paint dreams on the dying stone. Then later, much later, the ocean floor will be punctuated by Chevy trucks, carrying the dreamers' descendants, who are going to the store.

—Joy Harjo, *Invisible Fish*

Descending into a cave, eyes blink, thirsty for light. They absorb the low hues of earth as it becomes ceiling, walls, and floor; lungs inhale the mineral musk of damp soil. Distant pools murmur with the primordial stuff of life. *Worlds Below* unravels this elemental story: one of carbon, iron, nitrogen, even silver and gold, all formed from burning stars.

Earth writes its history in layers of rock. We can systematically dig it up or study banded mountains and canyon walls: time marked by strata of color and texture. The ancient deep is speckled with fossils of early life-forms. Much later, glass and metal tools appeared. Today, plastic detritus. Our footprint on time, once subtle, is increasingly invasive.

For those of us who stand above ground, spaces below can be intimidating, fearful. Our bodies cannot survive without sunlight and air. Yet the prospect of buried riches is seductive. We've hunted for treasure, resources, power, and energy, dredging ocean floors and boring deep into Earth.

Feats of technology enabled us to see, smash, and grab what before we could not. Scaffolding and tunnels plunge generations of miners, children included, down into darkness to scrape for precious metals or gems they cannot keep for themselves. In California alone, more than 45,000 now-abandoned mines haunt the landscape. We scour for glittering veins as well as darker gold. Coal for fuel. Or oil and gas, pulled upward by pumps, steel pipes, and enormous drills rigged on land and at sea. Perhaps soon, these will also litter the moon or Mars.

Below is ever enticing, ever transforming. Magma flows, seeds germinate, roots extend, elements crystallize; freshwater and saltwater carve fathomless networks. Time slows. And gravity pulls us, inevitably, down.

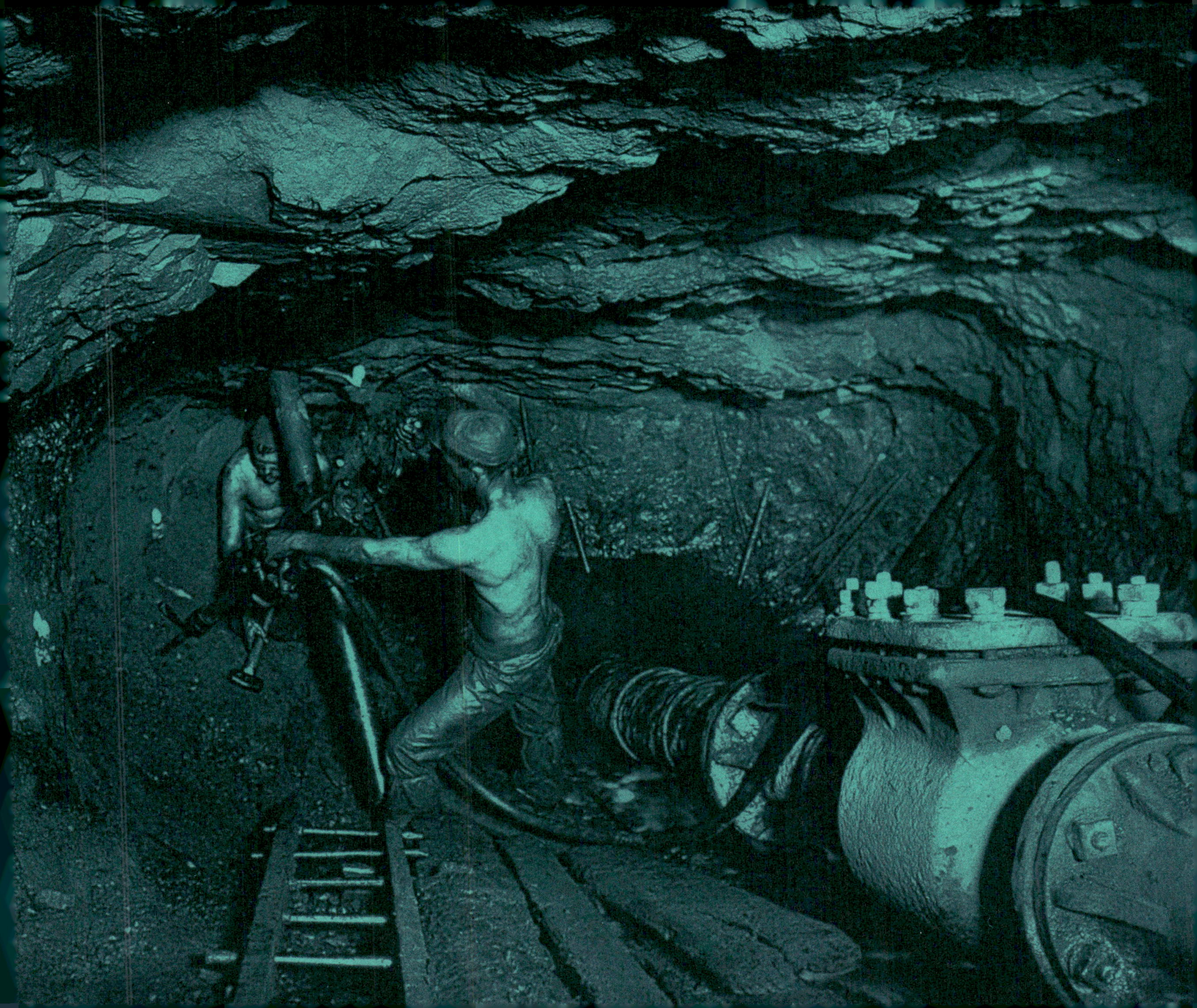

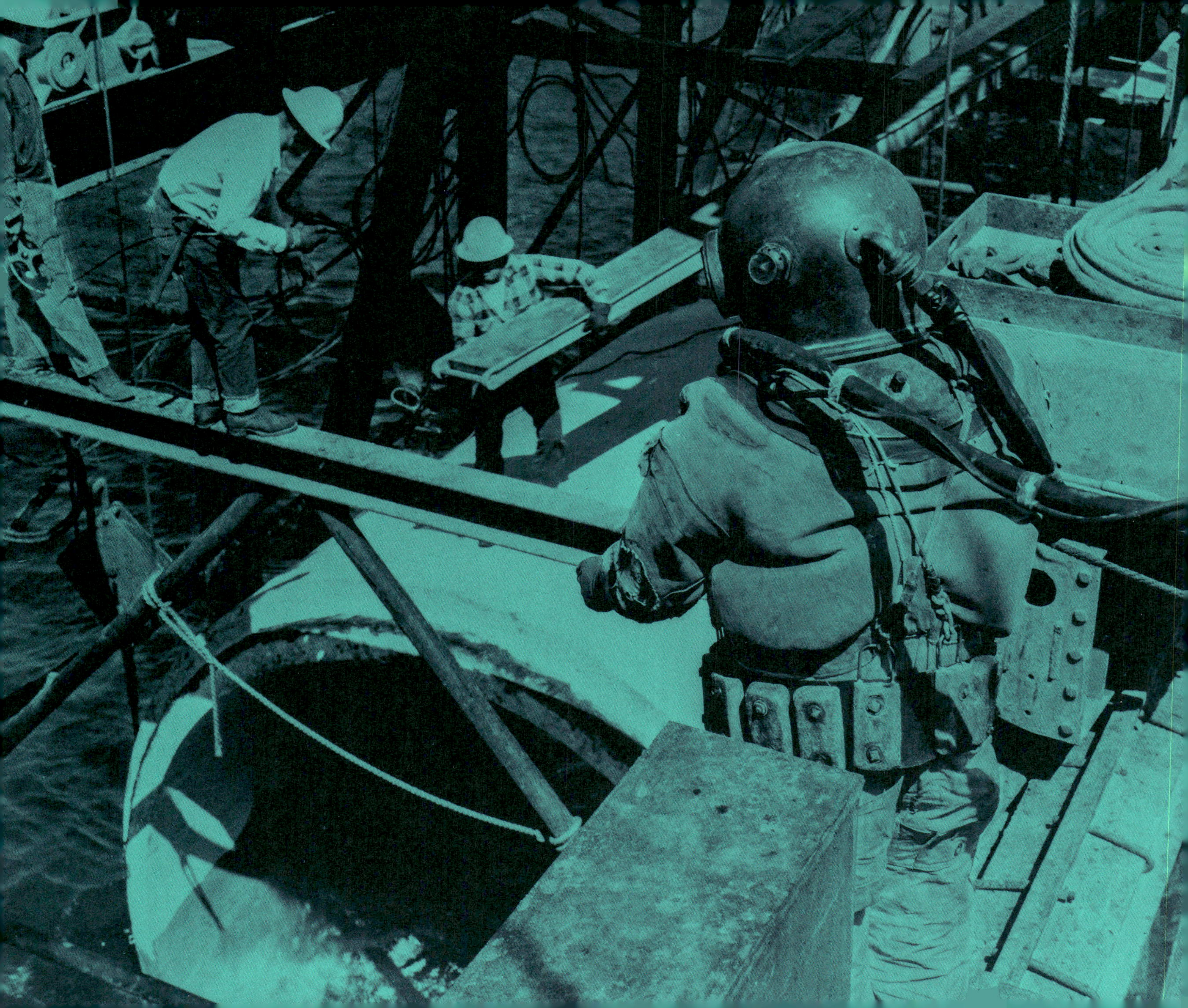

Fathom

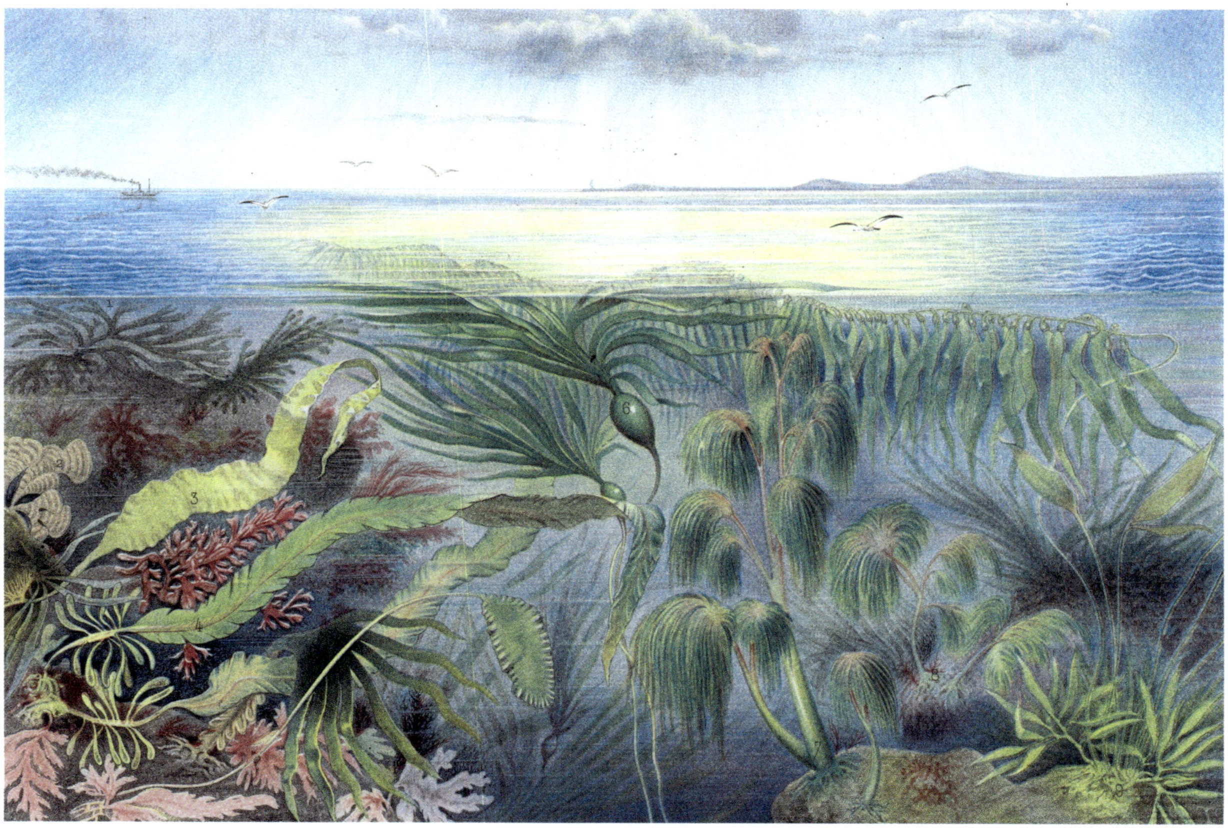

↓ *Mussel Diver, Thailand*, 1977.
Robert Glenn Ketchum (born 1947)

↩ ↘ *Das Meer* (The sea), 1867.
Matthias J. Schleiden (1804–1881)

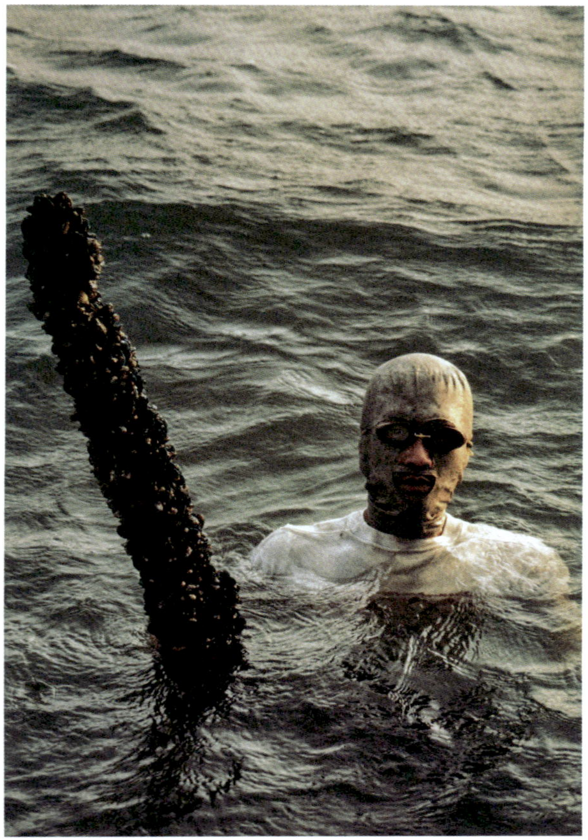

The Great Ocean

Water defines our planet. For centuries, communities in Southeast Asia and the Pacific Islands developed extensive knowledge of marine ecosystems as a source of food, 14 navigation, and spiritual connection. Much later, printed books began documenting the world below the waves. Two 17th-century English naturalists, Francis Willughby and John 30 Ray, are remembered for their laborious work classifying fish by physical traits like fins and gills. Their book contained 187 copperplate illustrations made by at least 18 artists. The German naturalist Marcus Bloch further organized this chaotic field with his encyclopedic approach, even describing fish he had never seen himself. In *Das Meer* (The sea), the German botanist Matthias Schleiden used engaging illustrations and prose to make marine biology and oceanography 91 accessible to general readers of all ages.

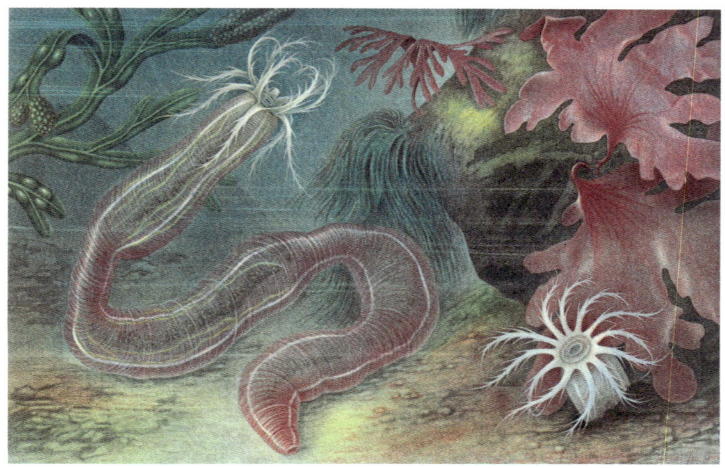

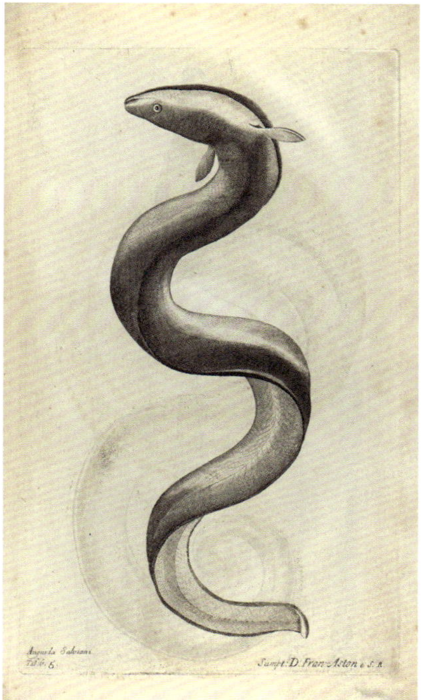

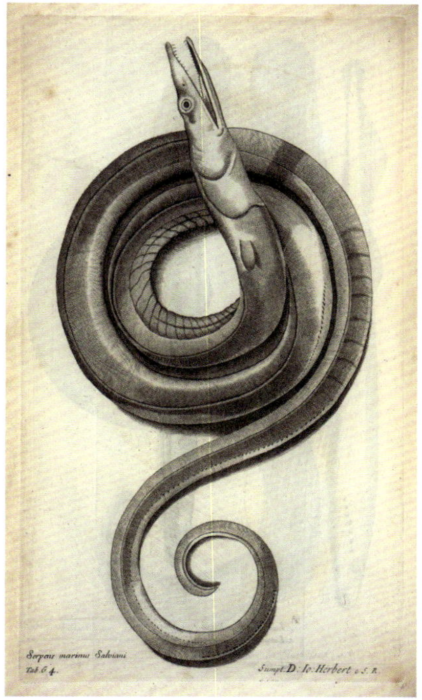

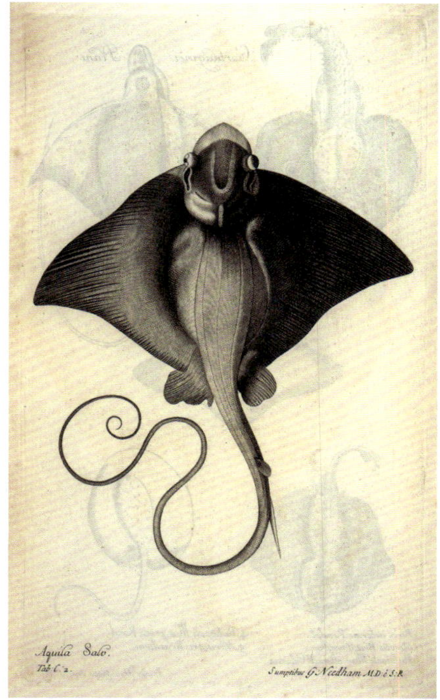

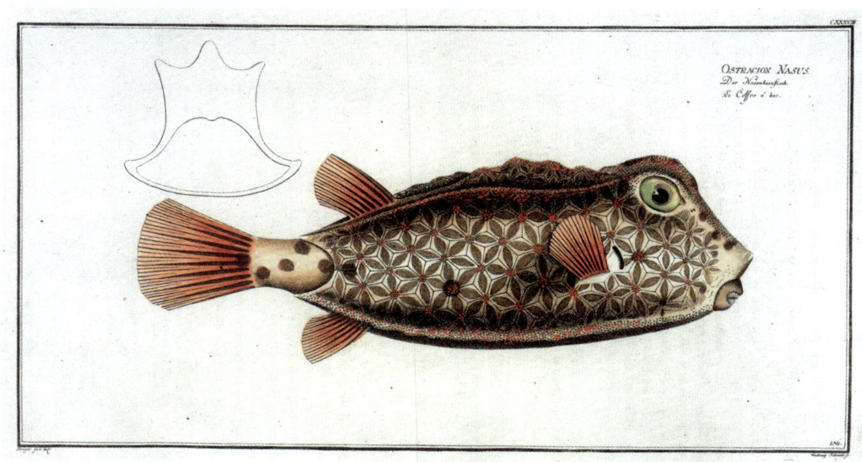

↑ *De historia piscium* (On the history of fish), 1686. Francis Willughby (1635–1672)

→ *Ichtyologie, ou, Histoire naturelle, générale et particulière des poissons* (Ichthyology, or, General and particular natural history of fishes), 1785–97. Marcus E. Bloch (1723–1799)

This is the first global view of our colorful, living biosphere. Published by NASA in the 1980s, it is a compilation of three years of data from two space satellites, visualizing patterns of plant growth on land and at sea. The equatorial region is notably abundant: Green areas mark dense forests, and violet indicates blooms of marine phytoplankton, or aquatic plant-like microorganisms (black areas indicate no data). Despite their size, phytoplankton create roughly half of our planet's oxygen and consume nearly 10 gigatons of carbon from the atmosphere each year. They are key to understanding how ocean ecosystems impact Earth's climate and temperature. ☽ 56

→ *Composite image of Earth's biosphere.*
NASA satellites NIMBUS-7 and NOAA-7

It is great. It terrifies, it frightens one. It is that which is irresistible; a great marvel; foaming, glistening, with waves; bitter—very bitter, most bitter; very salty. It has man-eating animals, animal life. It is that which surges. It stirs; it stretches ill-smelling, restless.

—Bernardino de Sahagún et al., Historia general de las cosas de Nueva España (Florentine Codex), ca. 1575–77

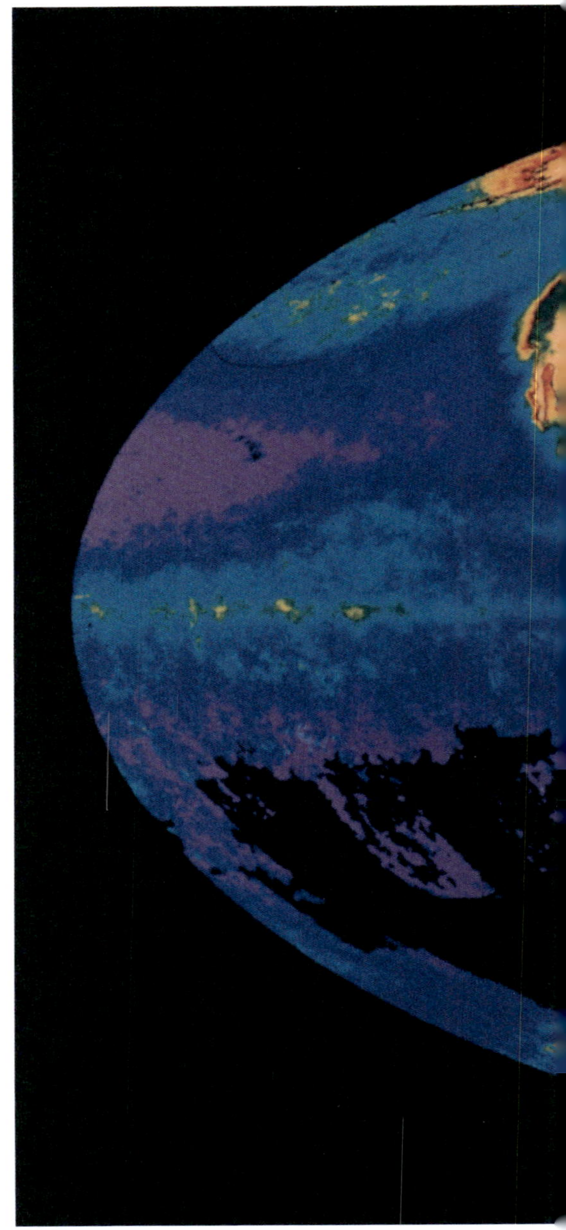

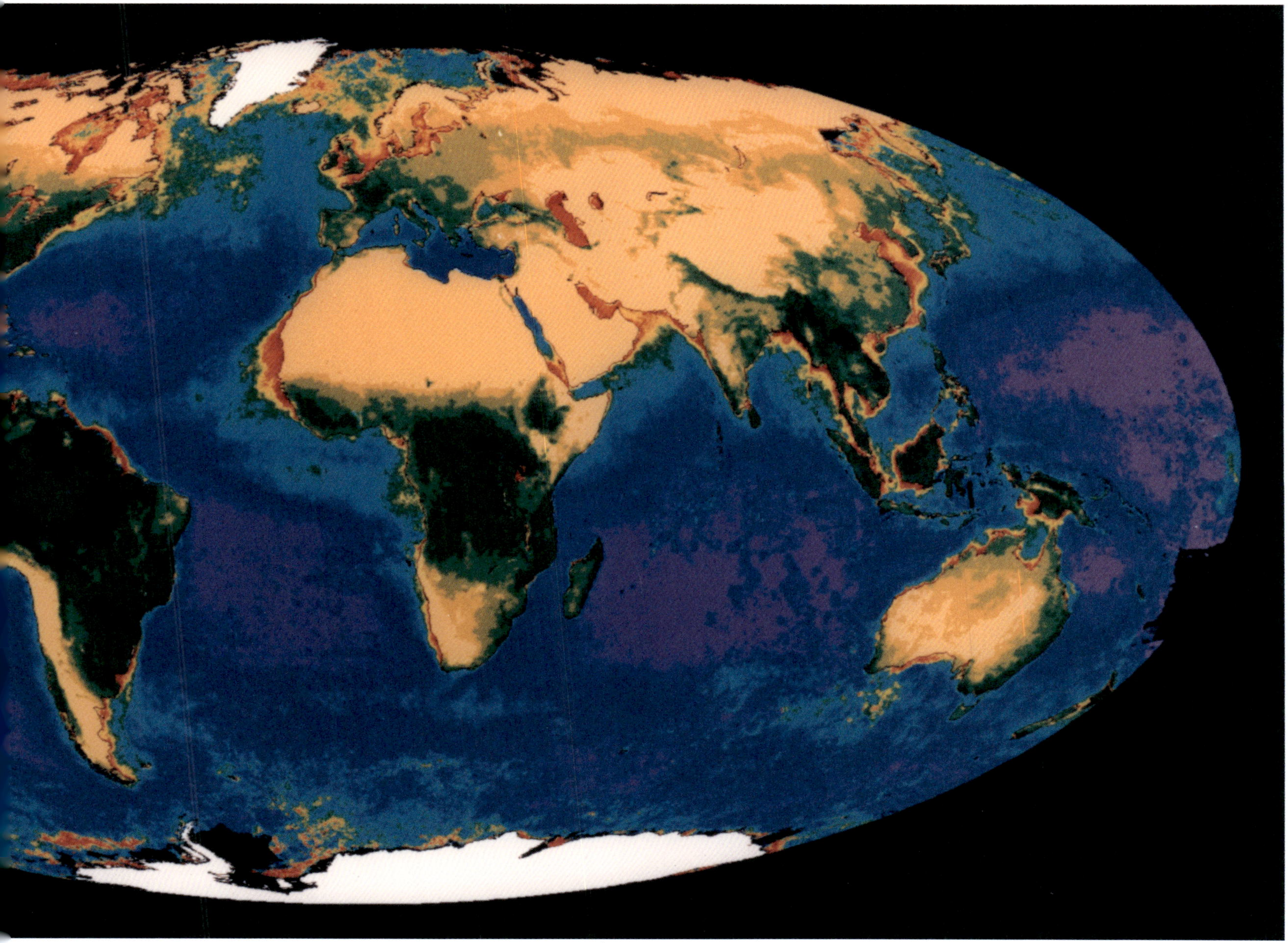

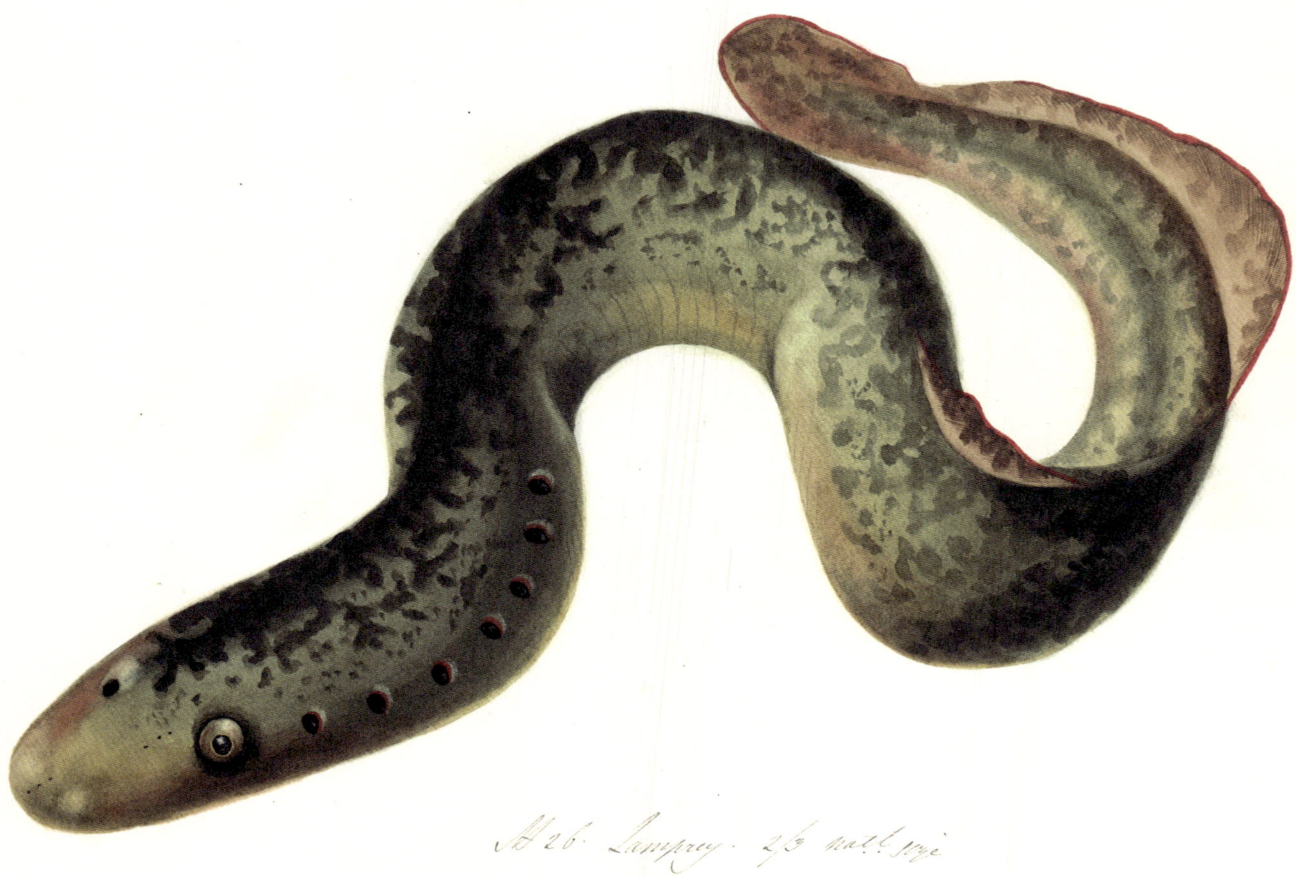
No 26. Lamprey. 2/3 natl. size

Freshwater Fish

The artist and author Sarah Bowdich spent 10 years crafting her field guide to Britain's freshwater fish. Every one of her publication's 47 illustrations is an original artwork, drawn and painted by Bowdich herself. She made a quick sketch of each fish immediately after it was caught, recording its colors, which changed dramatically with exposure to air. In the final paintings, she layered delicate, luminous tones, mimicking the prismatic surface of fish scales. Bowdich's text includes scientific identifications and her own observations. She kept pace with new research, and one of her later illustrations announced a long-awaited breakthrough: Naturalists had discovered how eels reproduce.

34–38

> The colours of the Trout change directly after they leave the stream; but I was lucky enough to avail myself of the skill of a friend, who supplied me with a succession of them as I sat on the bank, and by which I secured the tints in all their delicacy and brightness.
>
> —Sarah Bowdich Lee, *The Fresh-Water Fishes of Great Britain*, 1828

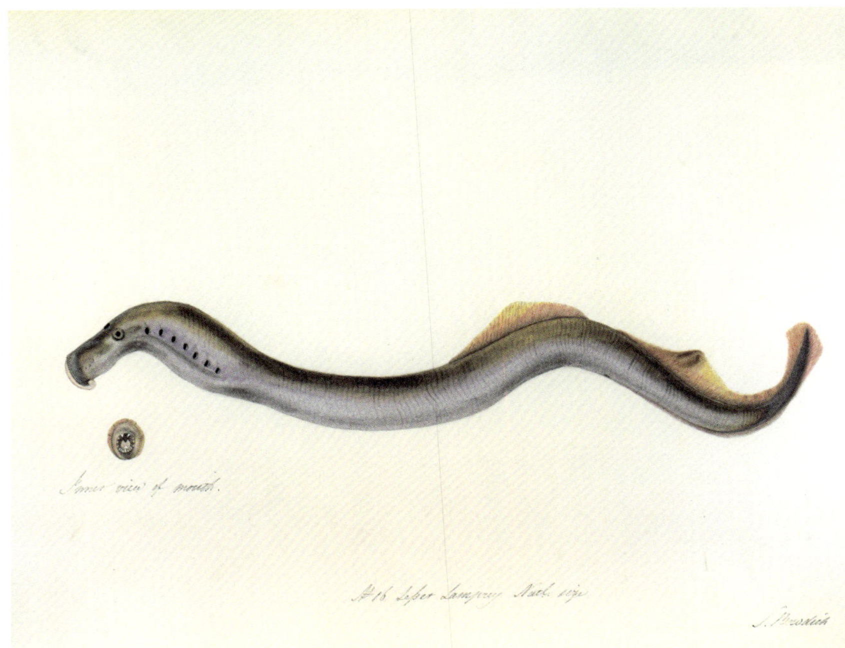

← → *The Fresh-Water Fishes of Great Britain*, 1828. Sarah Bowdich Lee (Mrs. R. Lee) (1791–1856)

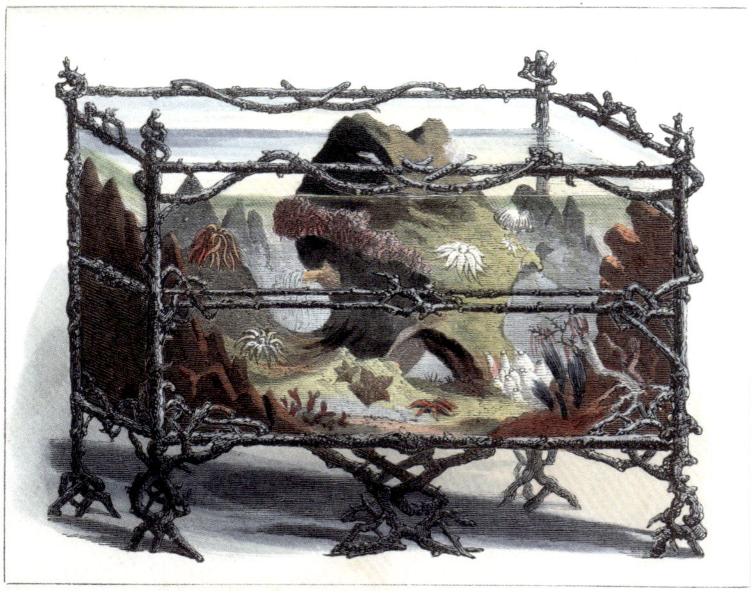

A DESIGN FOR AN AQUARIUM MOUNTED IN HANDSOME RUSTIC-WORK.

Wading In

In 19th-century Britain, searching for algae in tidal pools became a popular recreation for young and old alike. 42–45 Pressing these delicate specimens into albums blurred the line between botany and art. Some women, flouting Victorian gender roles, collected and studied seaweeds, 46 staking their claim to scientific work. The aquarium, a new form of scientific display and middle-class leisure activity, brought oceans and rivers indoors. The author Philip Gosse cultivated 75 species of sea anemones in his home, drawing them from life. "I have studied the living animals for eight years," he exulted, "which I describe with *distinctive precision* and *order*."

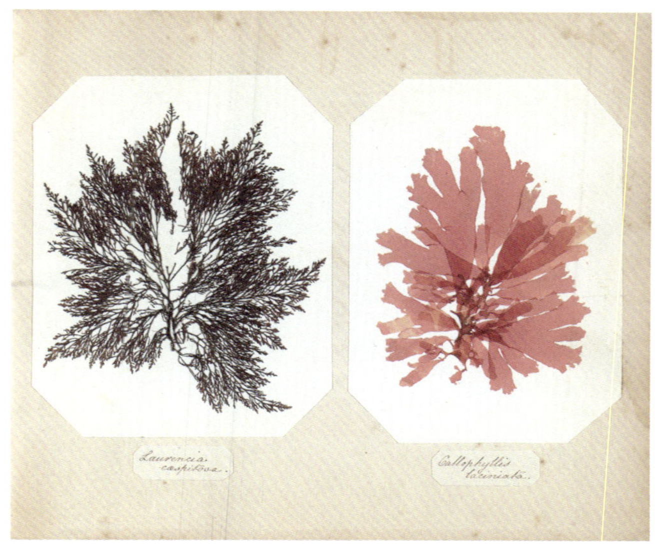

↑ *Ocean and River Gardens*, 1857. Henry Noel Humphreys (1810–1879)

→ Album of seaweed specimens. Collected, Arranged, and Mounted by Miss G. E. Dyke-Poore, Jersey, British Islands, 1855–85

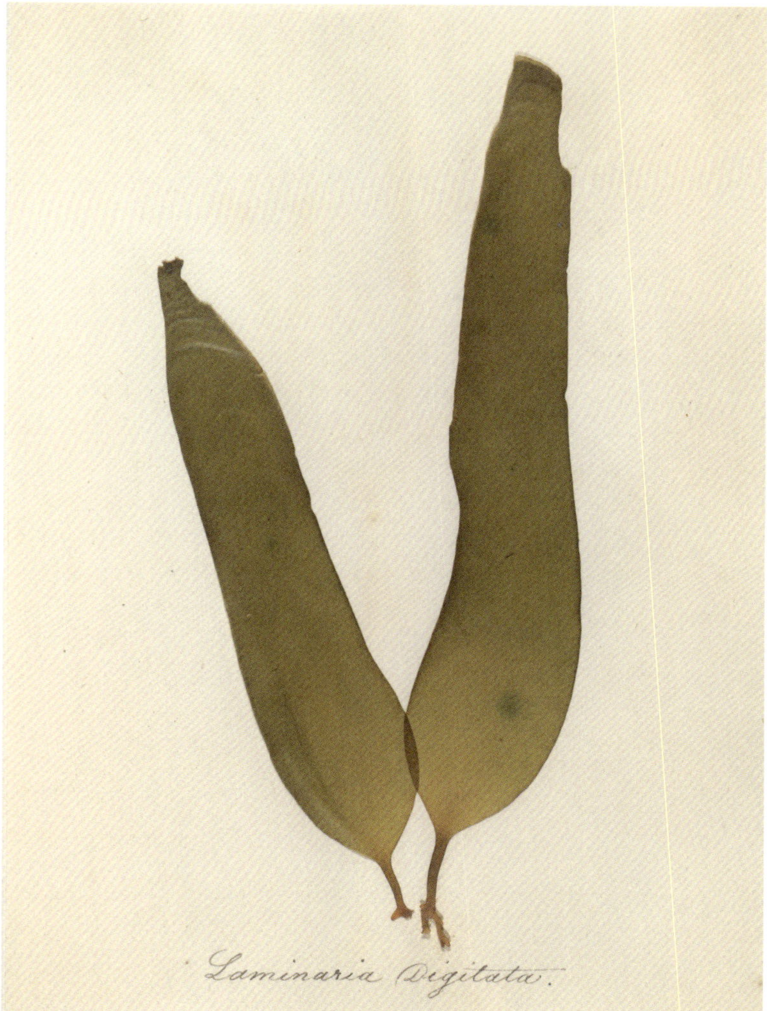

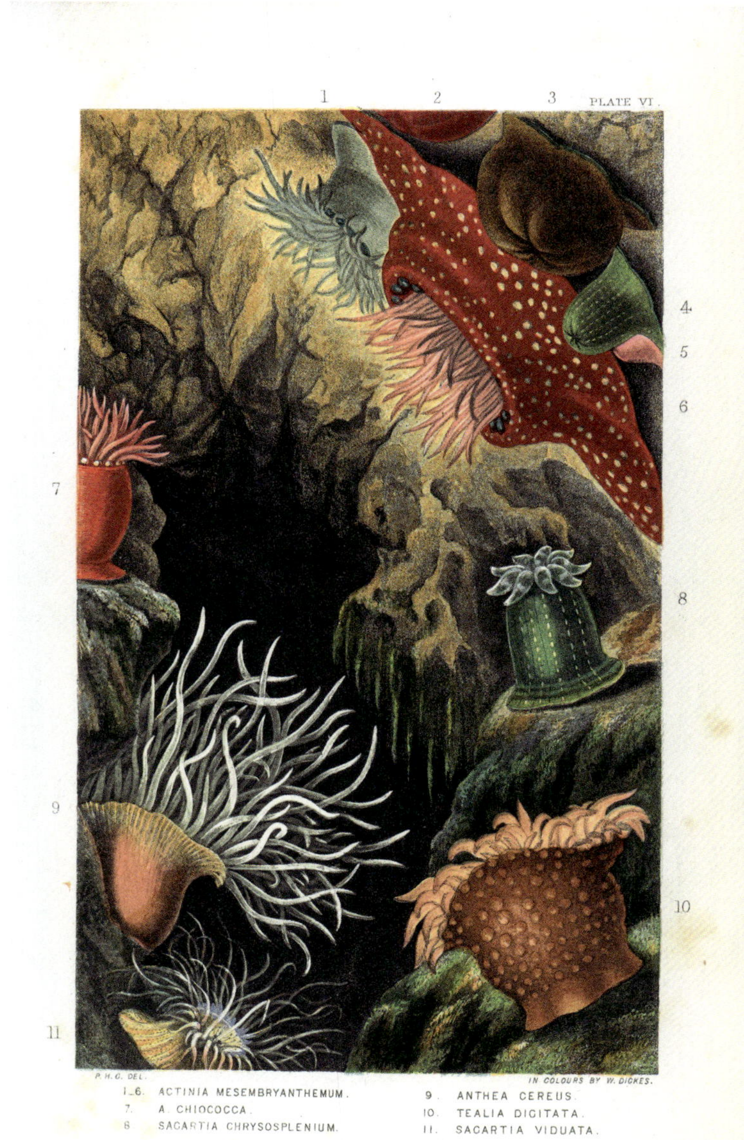

↑ Album of seaweed and algae specimens from the southern coast of England, ca. 1860

→ *Actinologia Britannica: A History of the British Sea-Anemones and Corals*, 1860. Philip H. Gosse (1810–1888), author; William Dickes (1815–1892), artist

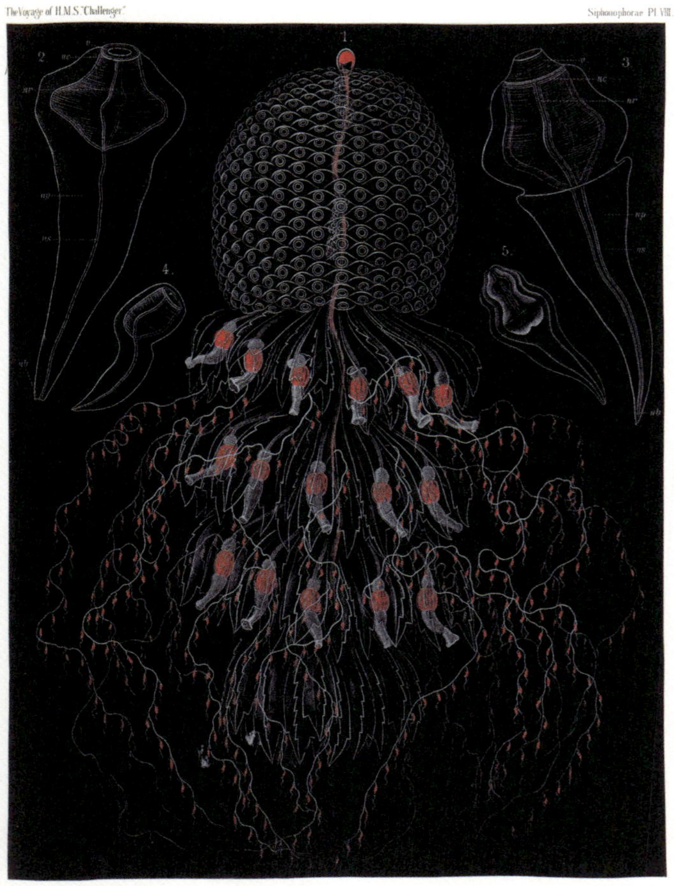

FORSKALIA THOLOIDES.

HMS Challenger

What exactly is the ocean floor composed of? Does life exist at the bottom of the sea? When HMS *Challenger* set sail in 1872, knowledge of Earth's largest ecosystem was patchy at 10 best. While most maritime expeditions mapped coastlines and nautical passageways, the *Challenger* studied the world below its hull. For three and a half years, a crew of sailors, instrument makers, naturalists, an artist, and a chemist circumnavigated the globe, collecting samples and measurements. Over the next 20 years, the *Challenger* team, based in Edinburgh, Scotland, sifted and interpreted a mountain of data, producing a 50-volume report. From red clay to oozing mud, plankton to sea pigs, corals, sharks, octopi, kelp, 4 fish, and more, the *Challenger* report revealed the oceans teeming with life. This sprawling endeavor is widely considered the foundation of modern oceanography.

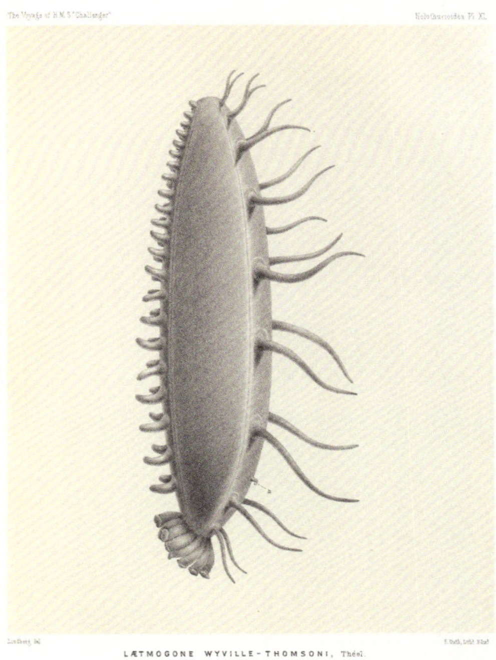

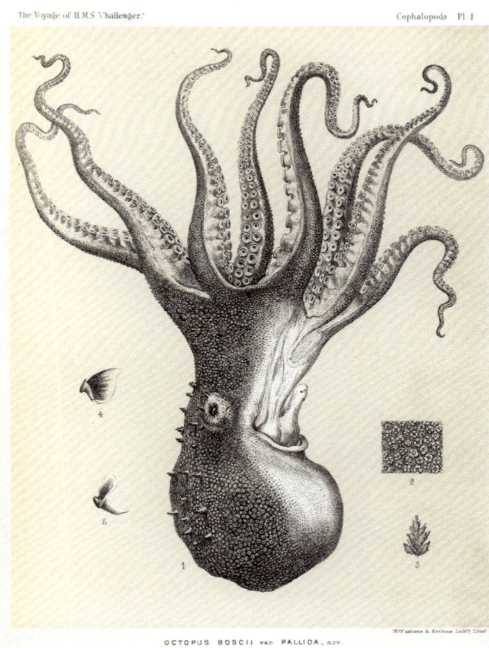

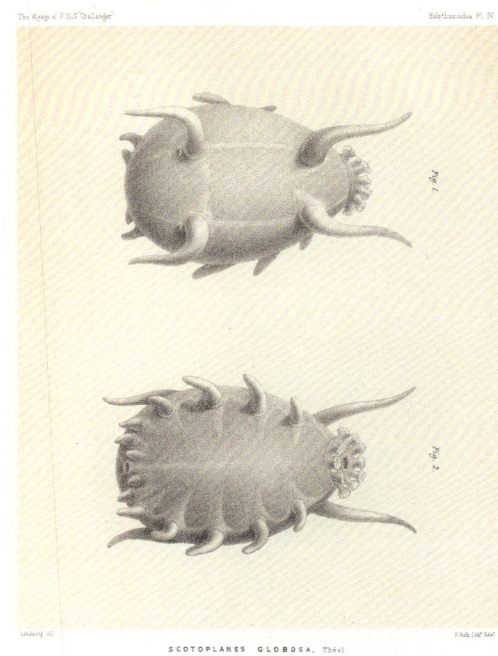

← ↑ *Report on the Scientific Results of the Voyage of HMS* Challenger *During the Years 1873–1876*, 1882–95. Charles Wyville Thomson (1830–1882) and John Murray (1841–1914)

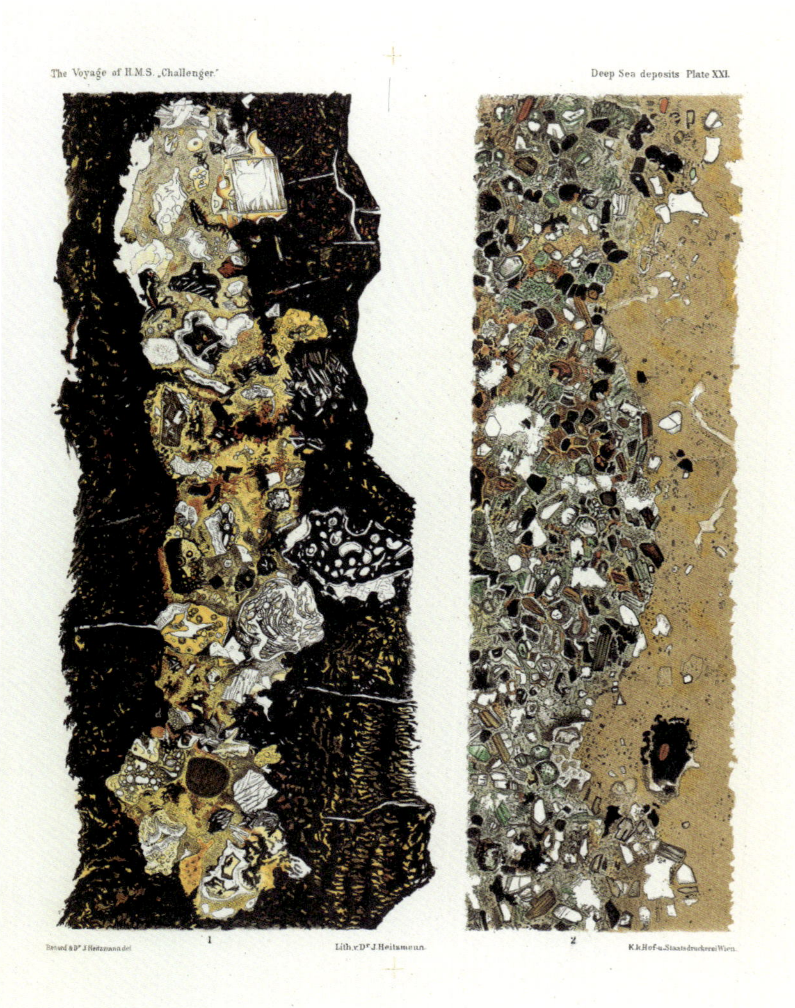

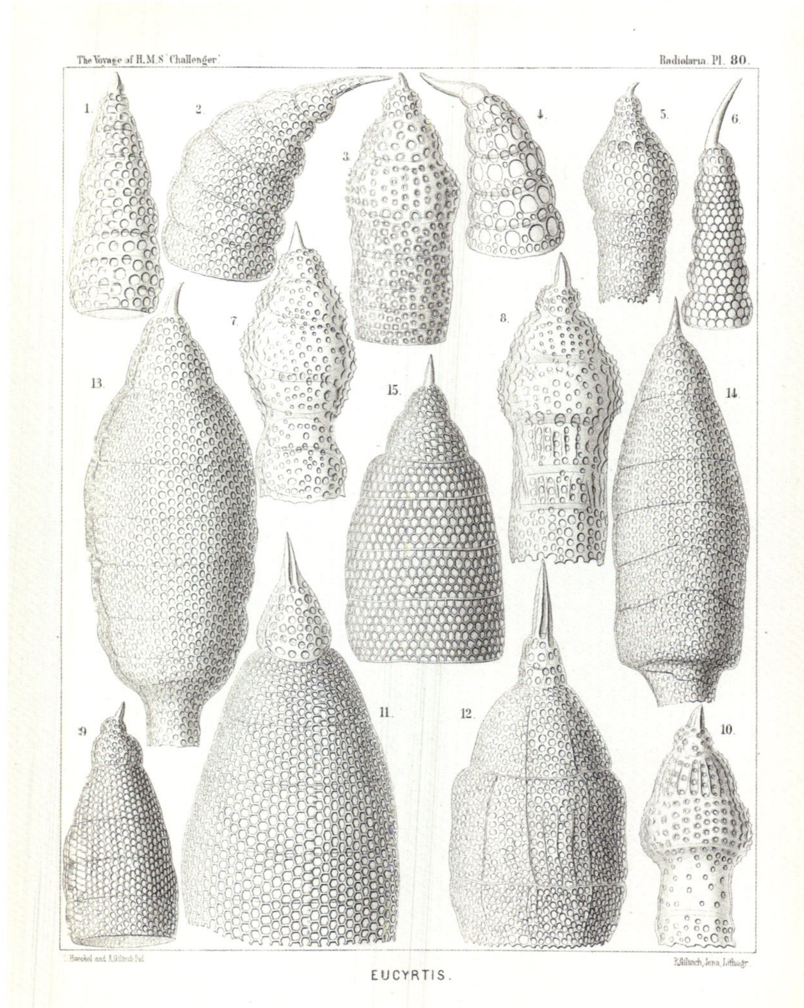

↑ → *Report on the Scientific Results of the Voyage of HMS* Challenger *During the Years 1873–1876*, 1882–95. Charles Wyville Thomson (1830–1882) and John Murray (1841–1914)

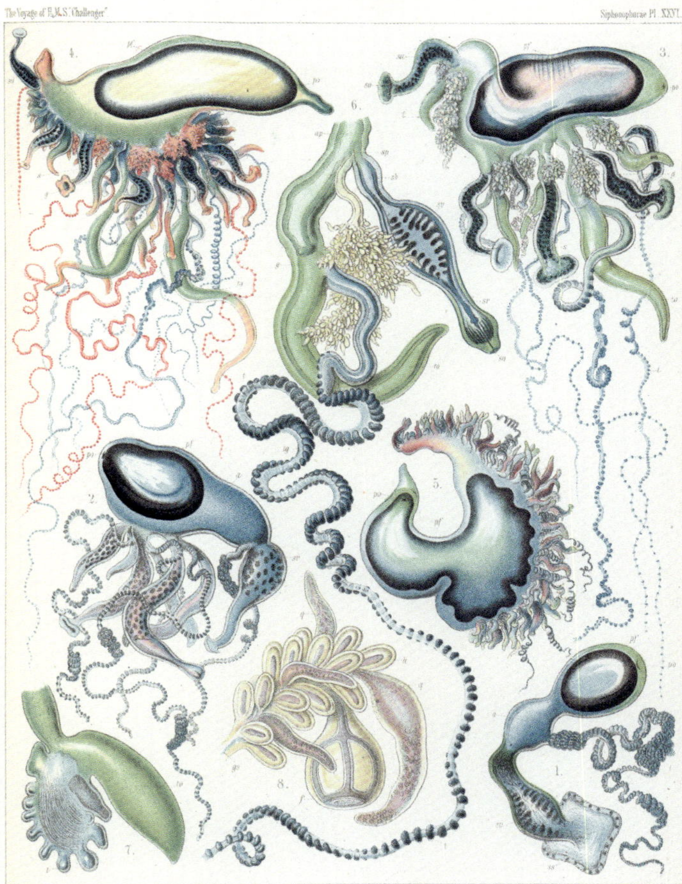

ALOPHOTA GILTSCHIANA.

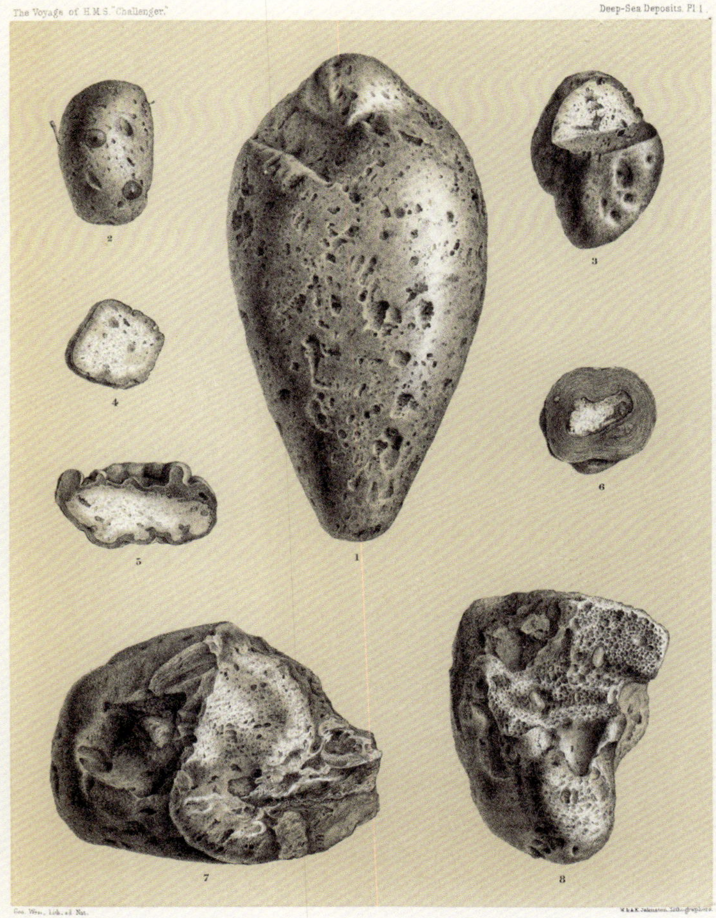

FRAGMENTS OF PUMICE AND MANGANESE-IRON NODULES.

"What an experience!" I exclaimed. "Traveling these deep regions where no man has ever ventured before! Look, captain! Look at these magnificent rocks, these uninhabited caves, these last global haunts where life is no longer possible!"
—Jules Verne, *Twenty Thousand Leagues under the Seas*, 1873

The deep-sea adventures of Captain Nemo in his submarine, the *Nautilus*, inspired scientists, artists, poets, philosophers, and students of all ages to dream about—and study—the fathoms below. The author Jules Verne first published *Vingt mille lieues sous les mers* as a magazine serial in his native France between 1869 and 1870. Submarines and deep-sea diving suits were still developing technologies, but Verne rendered imaginative and prescient descriptions of both. His protagonists witnessed a floating menagerie of marine life and even the transatlantic cable.

☽ 86–87

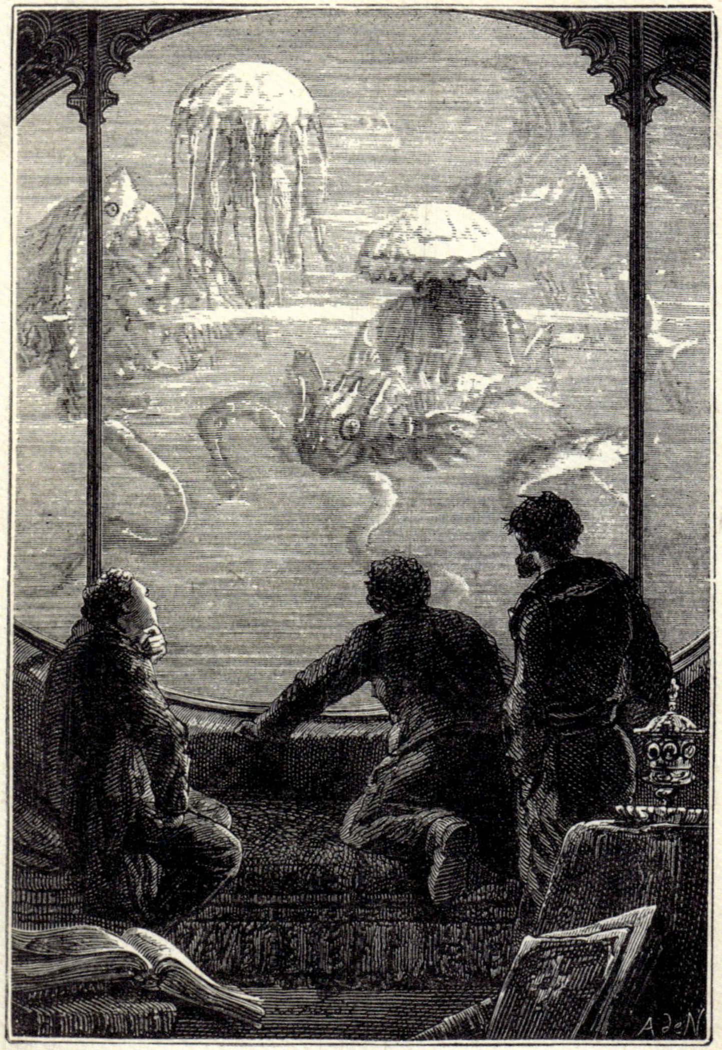

" A window opened into this unexplored abyss."—*Page* 76.

→ *Twenty Thousand Leagues under the Seas* (English edition), 1873. Jules Verne (1828–1905), author; Édouard Riou (1833–1900), artist

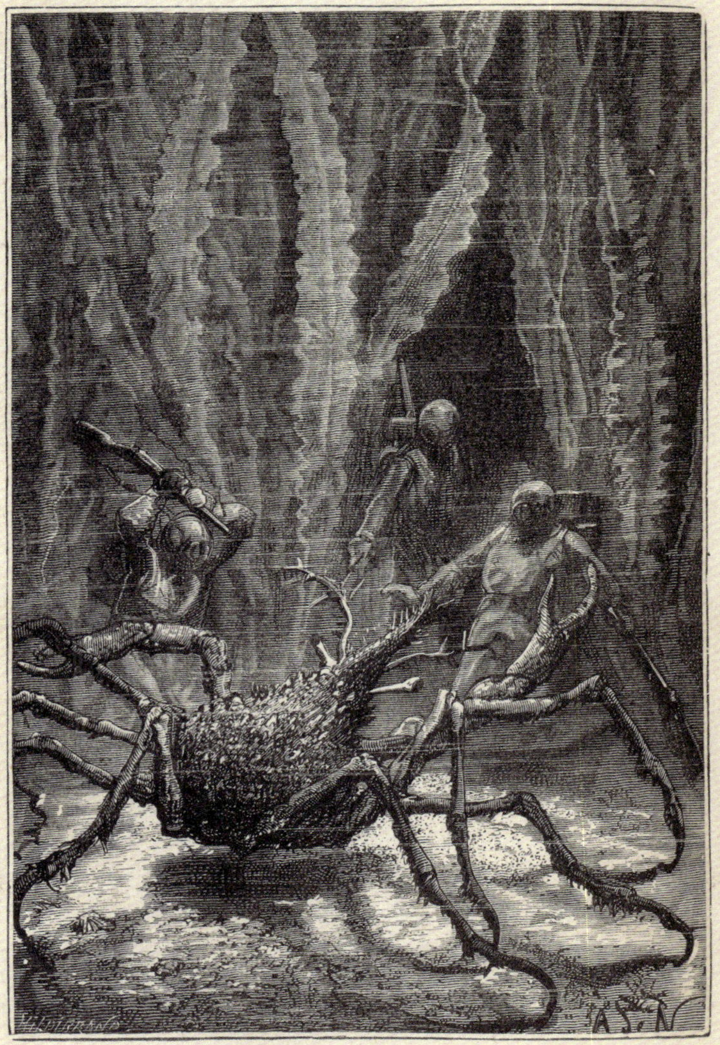

A monstrous sea-spider.—*Page 91.*

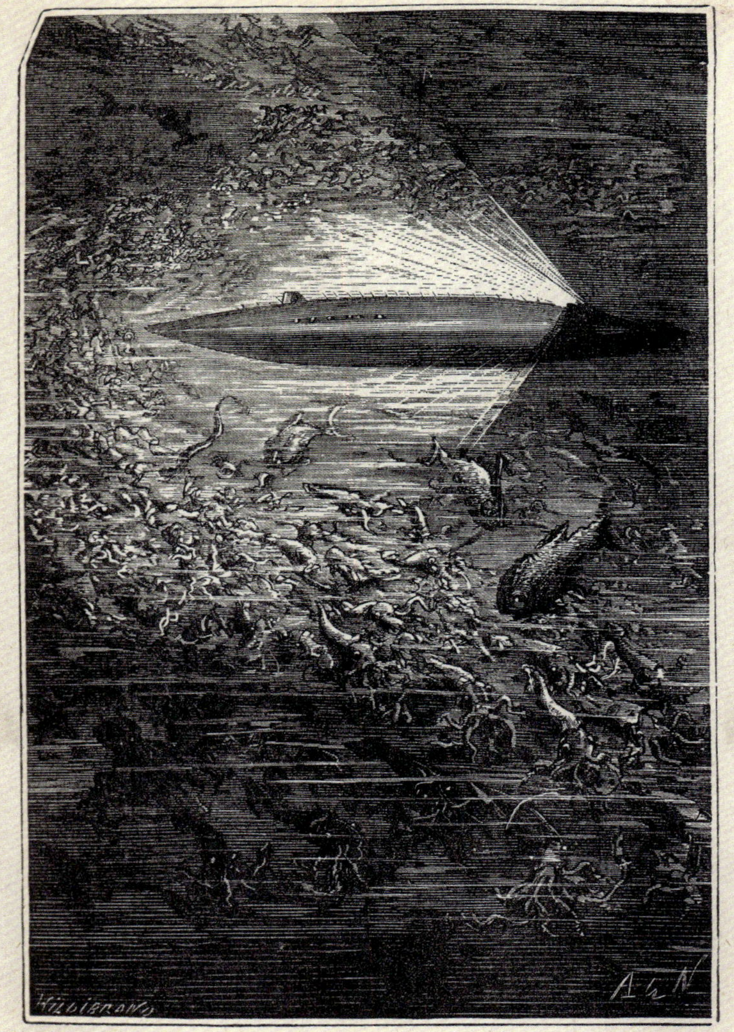

"One could count the calmars by millions."—*Page 98.*

Submarine Telegraphy

On one repair in mid-Atlantic, we hooked our cable twenty-six times in depths ranging from 2,000 to 2,600 fathoms, or about 15,000 feet. In each instance, we lifted the cable almost to the bow, only to see it snap under its own weight, and disappear into the sea.

—Captain William George Squares de Carteret, 1925

In 1866, the first successful submarine telegraph cable connected Ireland with Newfoundland. Three years later, a line linked Brest, France, to St. Pierre Island, off the Canadian coast. By the 1920s, more than 325,000 miles of copper wire allowed innumerable messages to skitter across the ocean floor. Deep-sea cables were sturdy: Their conductive copper core was insulated by layers of gutta percha (a natural latex harvested from Malaysian trees) and wound tightly by steel wire and hemp from the Philippines. The deepest lines were only one inch thick. Finding broken or faulty sections and then hooking, lifting, and rewiring them required extraordinary expertise. The most respected repair skipper, Captain William George Squares de Carteret, oversaw more than 400 mends. Cable clippings became commemorative items sold in velvet-lined boxes. Upon opening the glass lid, a powerful wave of tarry, briny odor wafts a signal from below.

41

24–25

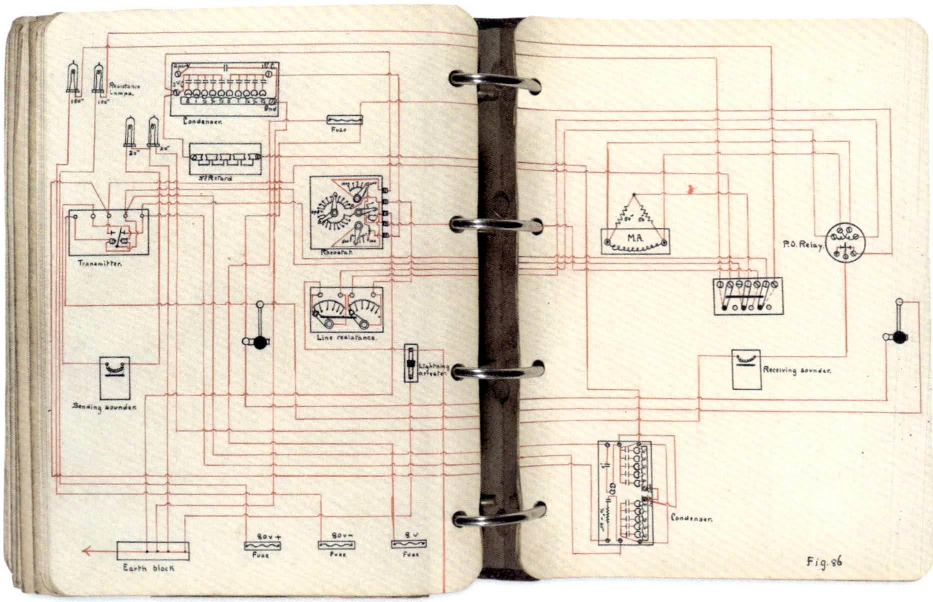

← → Notebook of engineer E. Pilkington, Western Union Cable Station, Bay Roberts, Newfoundland, ca. 1912–16

→ Box of telegraph cable laid in 1869 by Société du Câble Transatlantique Française (French Transatlantic Cable Society), from Brest to St. Pierre and St. Pierre to Duxbury, Massachusetts, ca. 19th century. Telegraph Construction & Maintenance Company

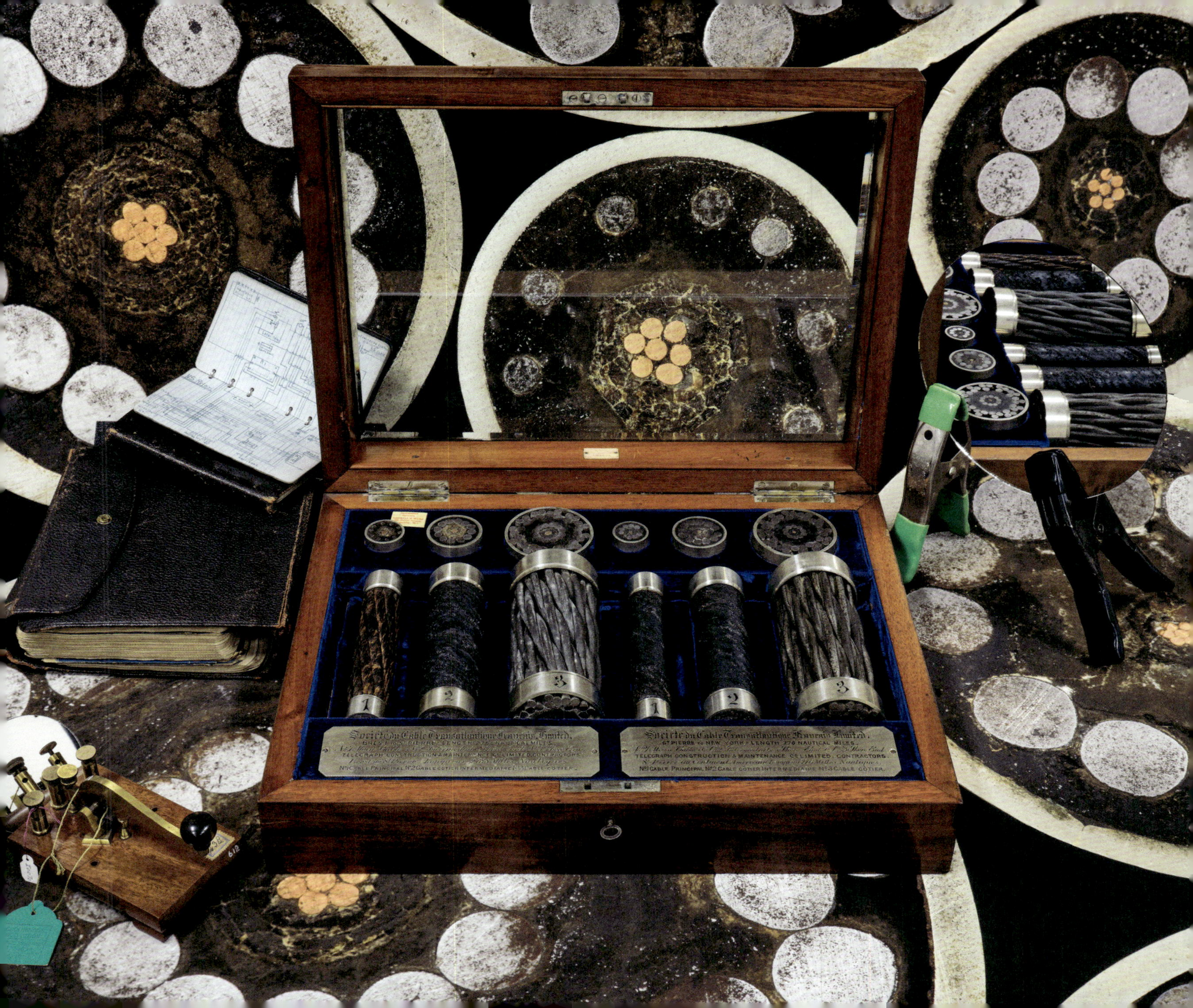
TELEGRAPH CONSTRUCTION & MAINTENANCE Co LIMITED, CONTRACTORS.
No1 CABLE PRINCIPAL No2 CABLE COTIER INTERMEDIAIRE No3 CABLE COTIER.

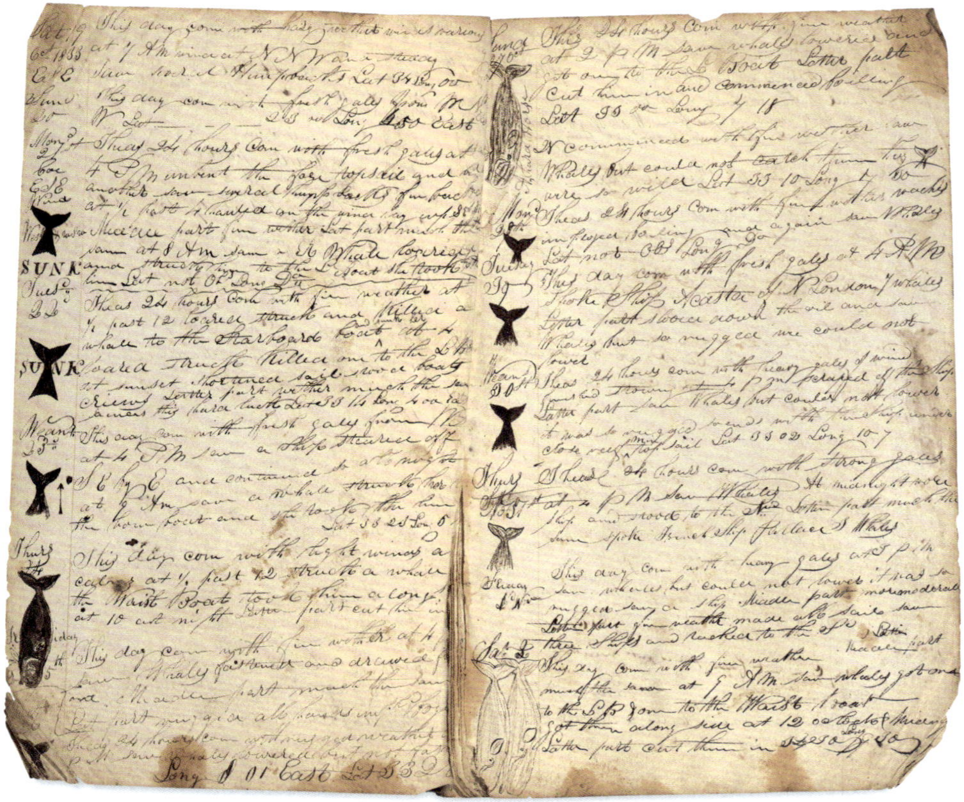

← Whaling voyage logbook, South Atlantic and the southwest Indian Oceans, 1833–35

↓ *A Most True and Marvelous Strange Wonder* broadside, 1568. Timothy Granger

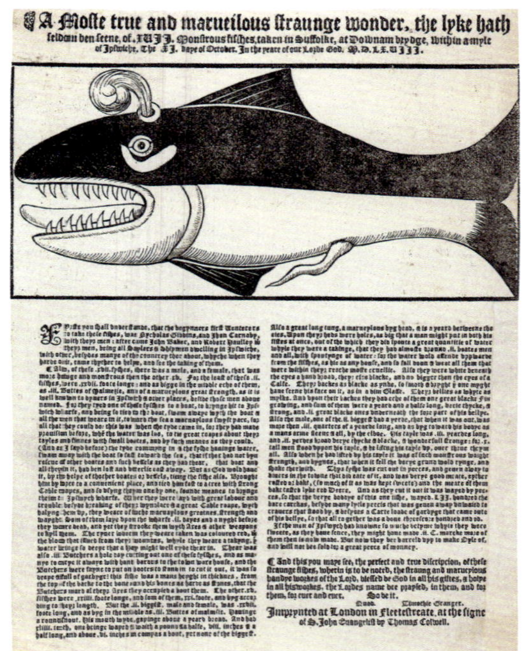

A Moste true and maruelous straunge wonder, the lyke hath seldom ben seene, of .XVII. Monstrous fisshes, taken in Suffolke, at Downam brydge, within a myle of Ipswiche. The .XI. daye of October. In the yeare of our Lorde God. M.D.LX.VIII.

And this you maye see, the perfect and true discripcion, of these straunge fishes, wherin is to be noted, the straung and maruelous handye workes of the Lord, blessed be God in all his giftes, & holye in all his workes, the Lordes name bee prayced, in them, and for them, for euer and euer. So be it.

Quod. Timothie Granger.

Imprynted at London in Fleetestreate, at the signe of S. John Euangelist by Thomas Colwell.

All the whales in the wider deeps, hot are they, as they urge on and on, and dive beneath the icebergs. The right whales, the sperm-whales, the hammer-heads, the killers there they blow, there they blow, hot wild white breath out of the sea!

—D.H. Lawrence, "Whales Weep Not!," 1932

Chasing Whales

A 1568 broadside recounts the brutal slaughter of a pod of orca whales that swam up the River Orwell toward Ipswich, England, giving an early glimpse into the fraught relationship between humans and Earth's largest mammals. In 1826, a Japanese text on farming proposes whale oil as a pesticide for crops infested with locusts. A few years later, handwritten logbooks and journals from sailors in the South Atlantic and Indian Oceans show whaling's increasingly global scale as demand for the mammals continued to grow, supplying baleen for corsets and umbrellas, and refined oils for candles, cosmetics, and industrial lubricants.

53

13

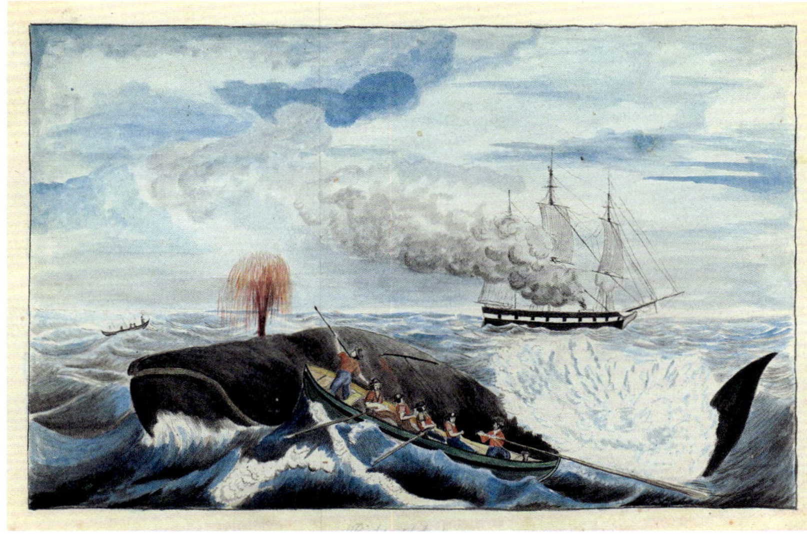

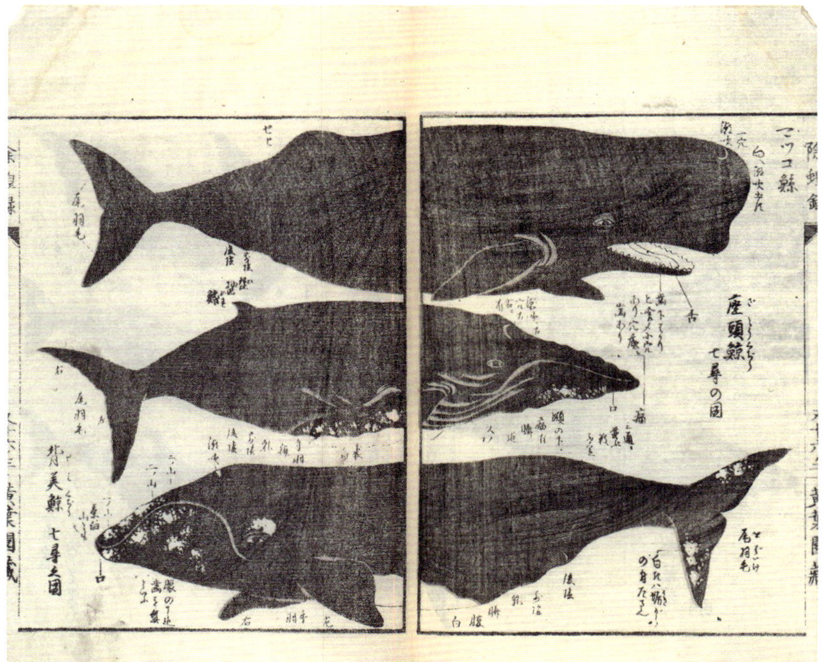

↗ (TOP) Color drawing of hunting right whales in ledger of whaling voyages, ca. 1851–57. David E. Marshall (born 1807); (BOTTOM) *Jokōroku* (Record of abolishing locusts), 1826. Ōkura Nagatsune (1768–1860)

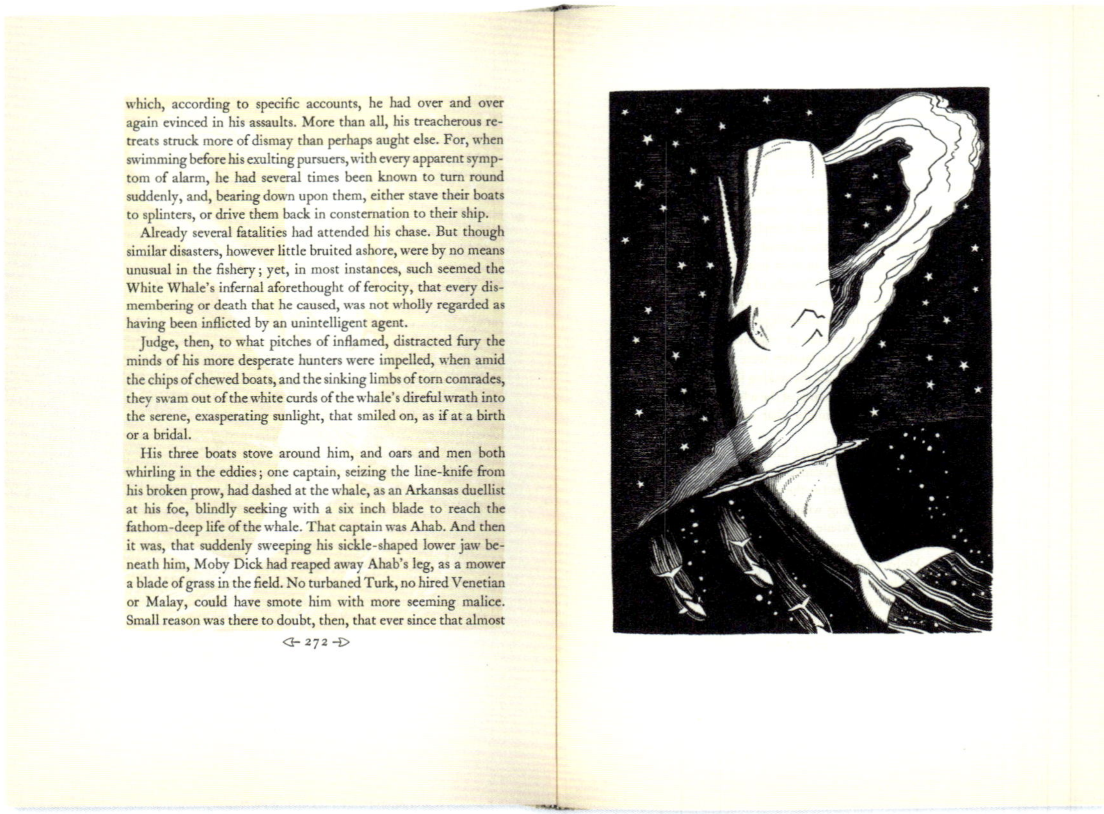

which, according to specific accounts, he had over and over again evinced in his assaults. More than all, his treacherous retreats struck more of dismay than perhaps aught else. For, when swimming before his exulting pursuers, with every apparent symptom of alarm, he had several times been known to turn round suddenly, and, bearing down upon them, either stave their boats to splinters, or drive them back in consternation to their ship.

Already several fatalities had attended his chase. But though similar disasters, however little bruited ashore, were by no means unusual in the fishery; yet, in most instances, such seemed the White Whale's infernal aforethought of ferocity, that every dismembering or death that he caused, was not wholly regarded as having been inflicted by an unintelligent agent.

Judge, then, to what pitches of inflamed, distracted fury the minds of his more desperate hunters were impelled, when amid the chips of chewed boats, and the sinking limbs of torn comrades, they swam out of the white curds of the whale's direful wrath into the serene, exasperating sunlight, that smiled on, as if at a birth or a bridal.

His three boats stove around him, and oars and men both whirling in the eddies; one captain, seizing the line-knife from his broken prow, had dashed at the whale, as an Arkansas duellist at his foe, blindly seeking with a six inch blade to reach the fathom-deep life of the whale. That captain was Ahab. And then it was, that suddenly sweeping his sickle-shaped lower jaw beneath him, Moby Dick had reaped away Ahab's leg, as a mower a blade of grass in the field. No turbaned Turk, no hired Venetian or Malay, could have smote him with more seeming malice. Small reason was there to doubt, then, that ever since that almost

272

The White Whale; a name, indeed, literally justified by his vivid aspect, when seen gliding at high noon through a dark blue sea, leaving a milky-way wake of creamy foam, all spangled with golden gleamings.

—Herman Melville, *Moby-Dick*

↑ *Moby Dick, or, the Whale*, 1851 (1930 edition). Herman Melville (1819–1891), author; Rockwell Kent (1882–1971), artist

Captain Ahab, monomaniacal master of the *Pequod*, a Nantucket whaling ship, sails across the world seeking revenge. Moby-Dick, the titular white whale, is the focus of the captain's ire as—across 600 pages—Ahab becomes increasingly unhinged. A classic of American literature, the book is a primer on every aspect of 19th-century whaling, whales, and human nature unleashed in a wild, watery world.

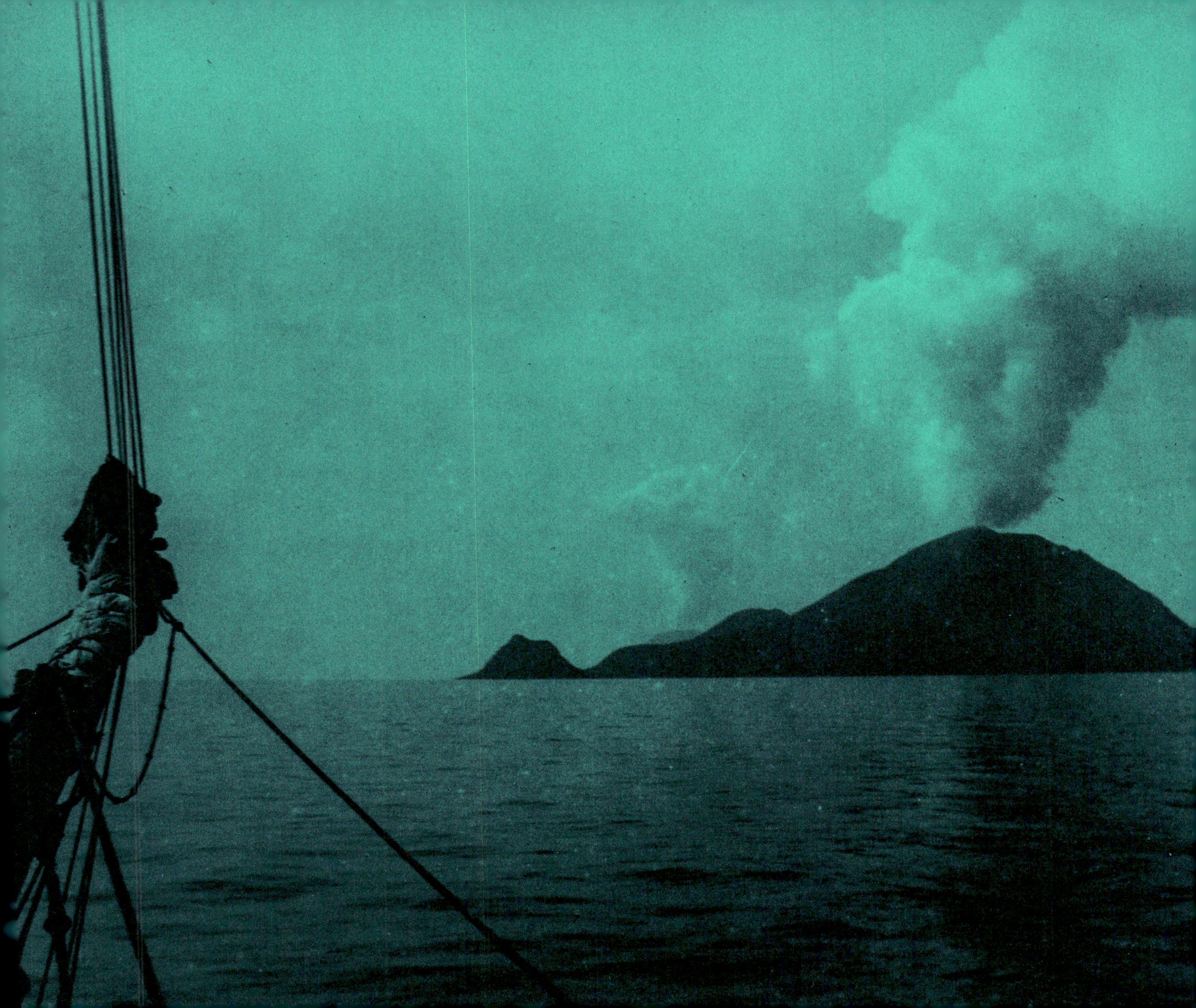

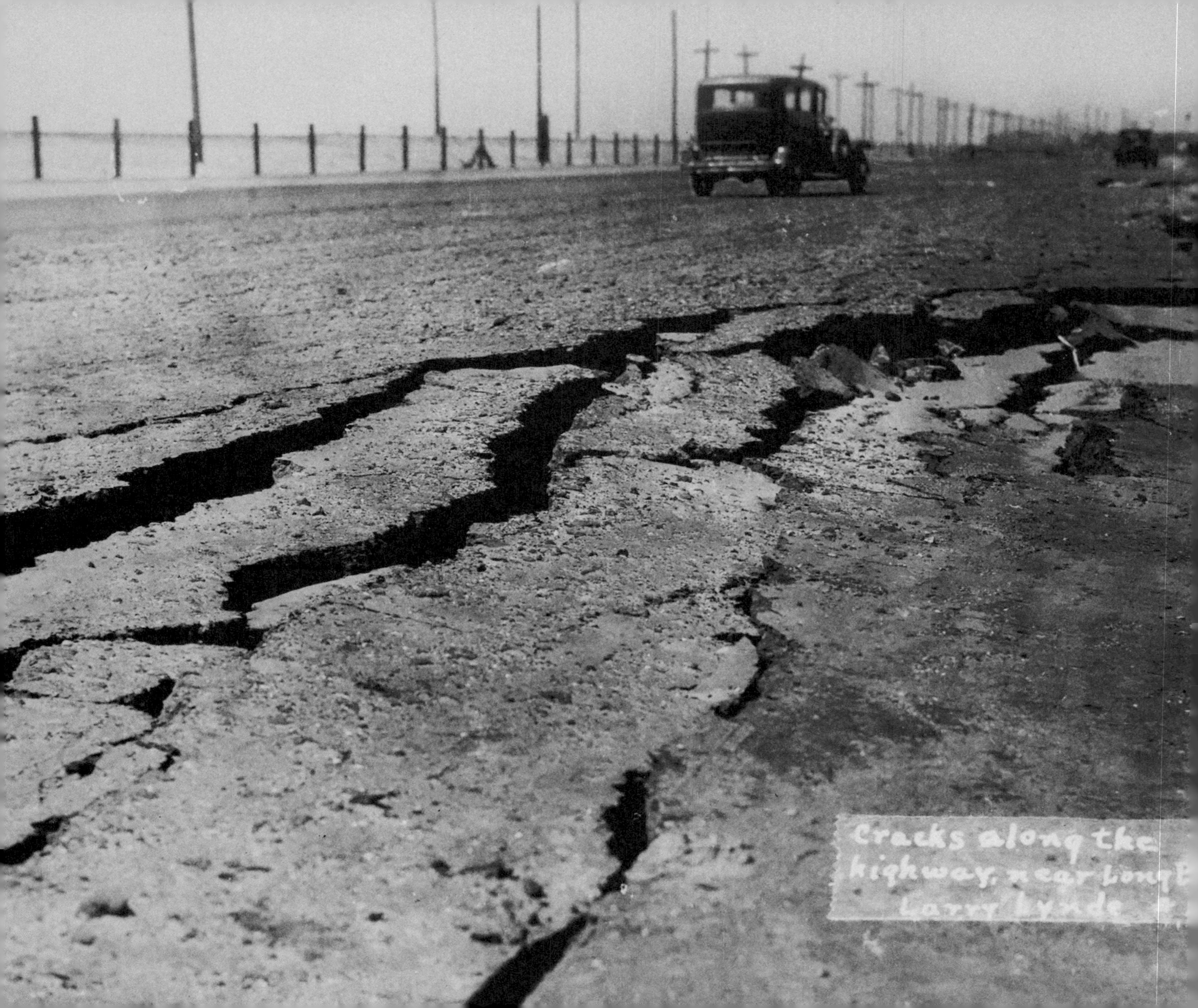
Cracks along the
highway, near Long
Larry Lynde

Rupture

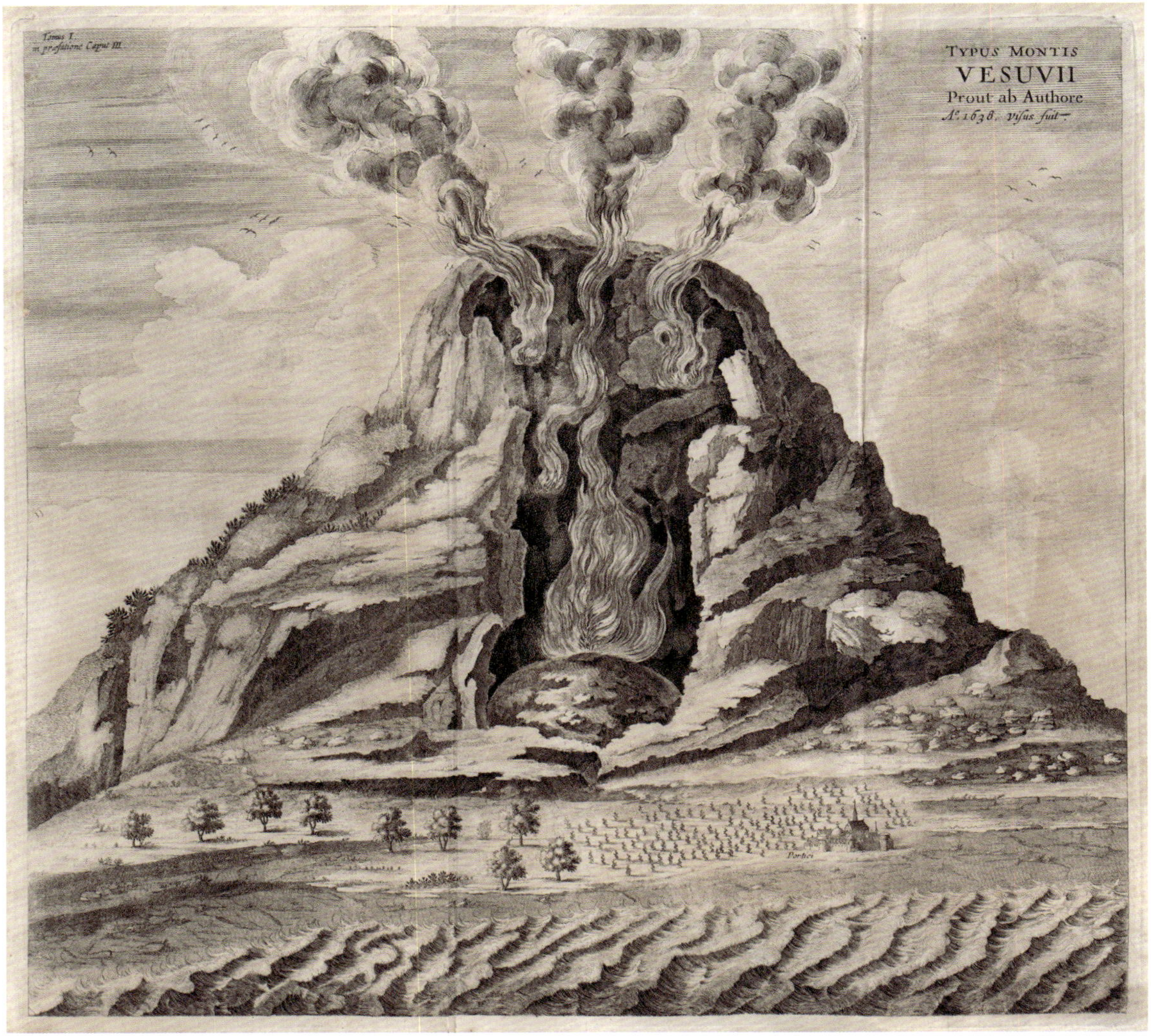
Tomus I.
in præfatione Caput III.
TYPUS MONTIS
VESUVII
Prout ab Authore
A°. 1638. visus fuit.

Fault Lines

California is seismically notorious. Through the centuries, temblors struck the region with virtually no warning, reshaping landscapes and upending lives. City planning changed dramatically because of such quakes, with revisions to building codes, zoning laws, and infrastructure design. In the 1990s, Caltech and the United States Geological Survey developed ShakeMap, designed to map the intensity of earthquakes across the world. Years of development then led to the creation of ShakeAlert, which currently provides people on the West Coast with precious time—up to 45 seconds—before a temblor strikes. Southern California experiences about 10,000 earthquakes per year, though most are too small to be felt.

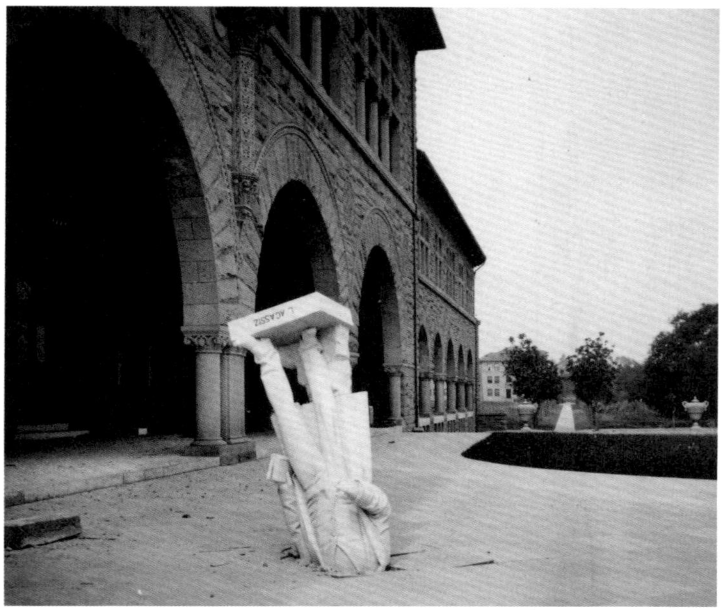

← *Mundus subterraneus* (Subterranean world), 1669. Athanasius Kircher (1602–1680)

← *Louis Agassiz statue at Stanford University*, April 21, 1906. W. C. Mendenhall

→ (LEFT) *Apartment and Store, Anaheim Blvd., Long Beach*, 1933. Austin Studio; (RIGHT) *Residence on Elm, Long Beach*, 1933.

↓ *Bird's-eye view of ruins of San Francisco from Captive Airship 600 feet above Folsom between Fifth and Sixth*, 1906. George R. Lawrence (1868–1938)

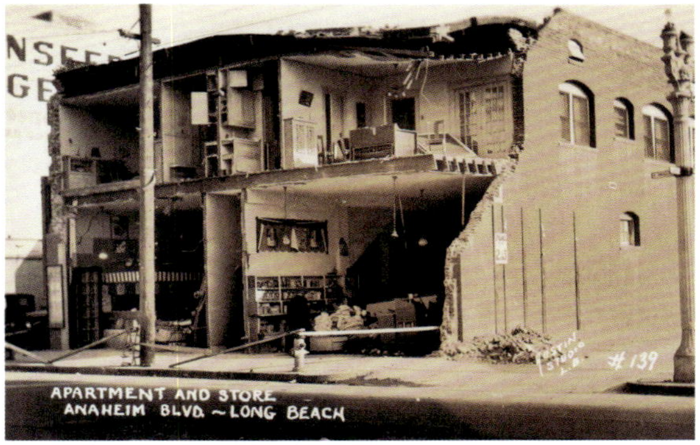

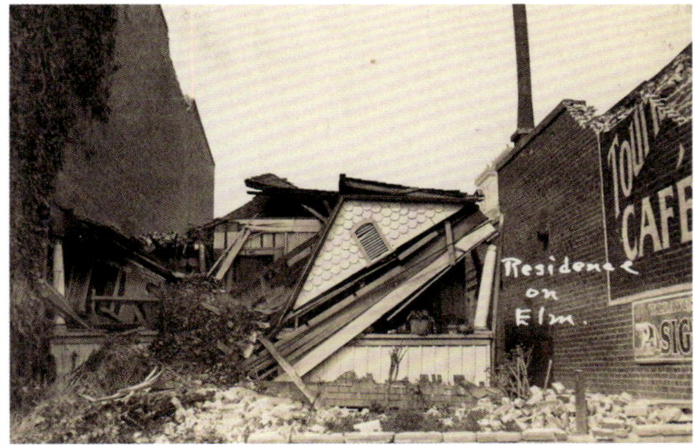

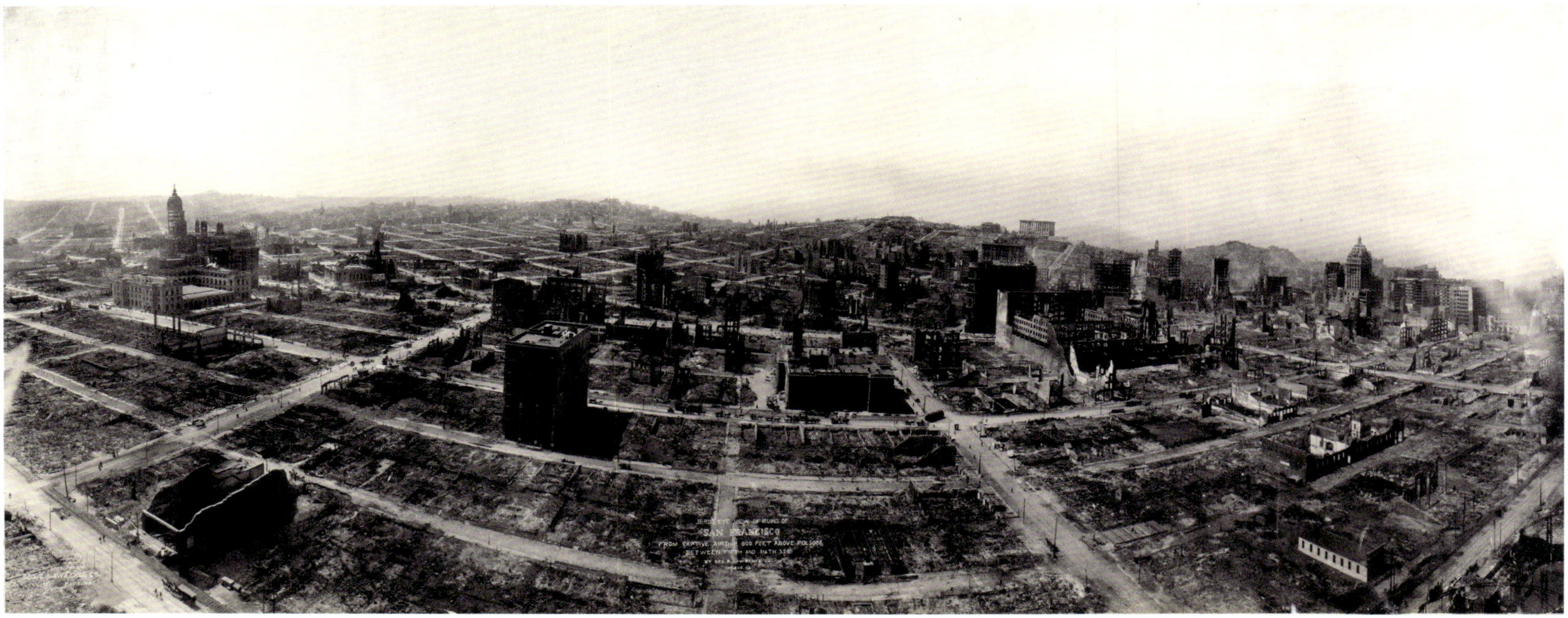

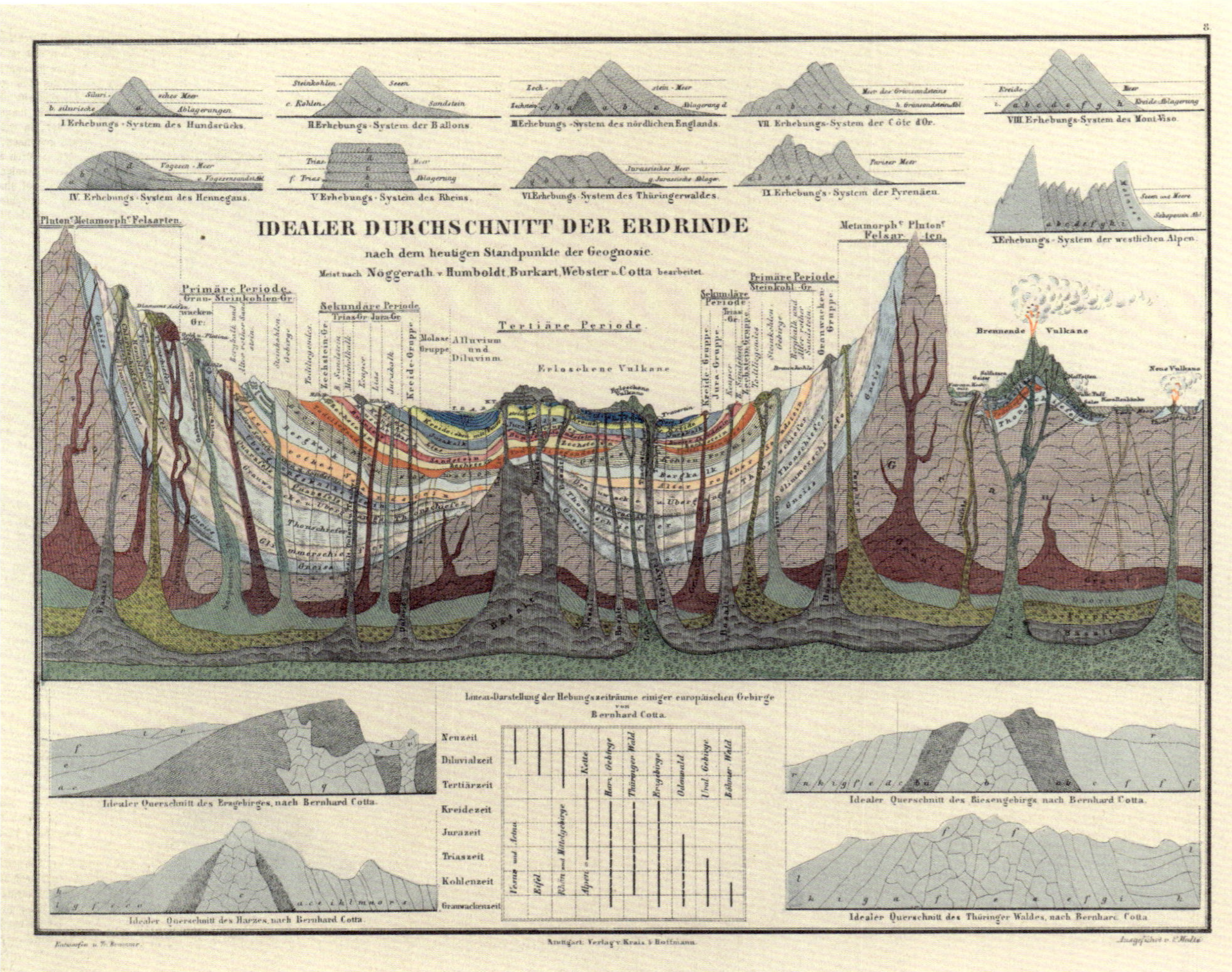

↑ Visualization of Earth's crust in *Atlas zu Alex. v. Humboldt's Kosmos in 42 Colorirten Tafeln* (Atlas to Alex. von Humboldt's Cosmos in 42 colored plates), 1853. Alexander von Humboldt (1769–1859), author; Traugott Bromme (1802–1866), engraver

Subterranean Fire

Volcanic eruptions inspire awe, fear, and questions about mysteries below ground. In 1665, the Jesuit polymath Athanasius Kircher imagined a volatile interior world, with volcanoes as vents for Earth's inner fires. 31, 37, BACK FLAP In the 18th century, Sir William Hamilton, known as the "volcano man," turned Vesuvius into a theater of observation, documenting its eruptions 56 in text accompanied by vivid, hand-colored plates. The explorer Alexander von Humboldt climbed the Andes and compared peaks across continents, arguing that they were part of a connected system shaped by climate, elevation, and dynamic geological forces. An image of lava bubbles in Kalapana, Hawai'i, documents study of Earth's active volcanoes as models for understanding those on other planets. 90

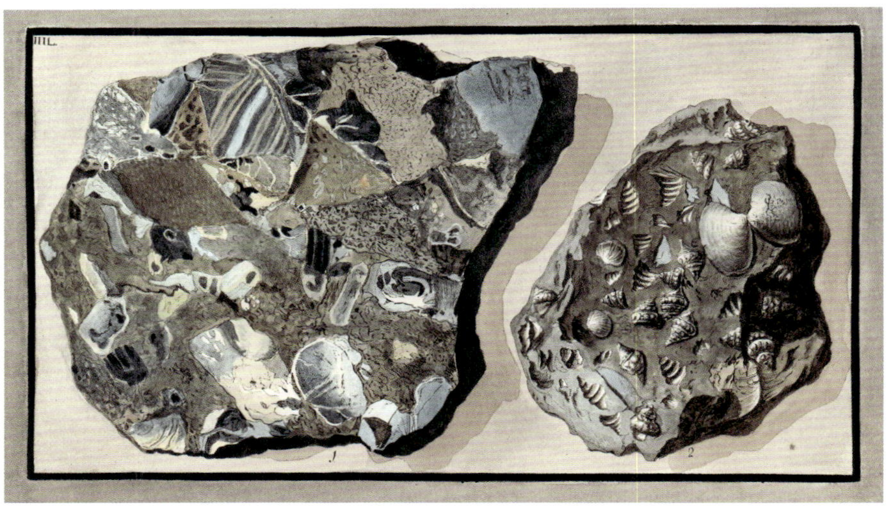

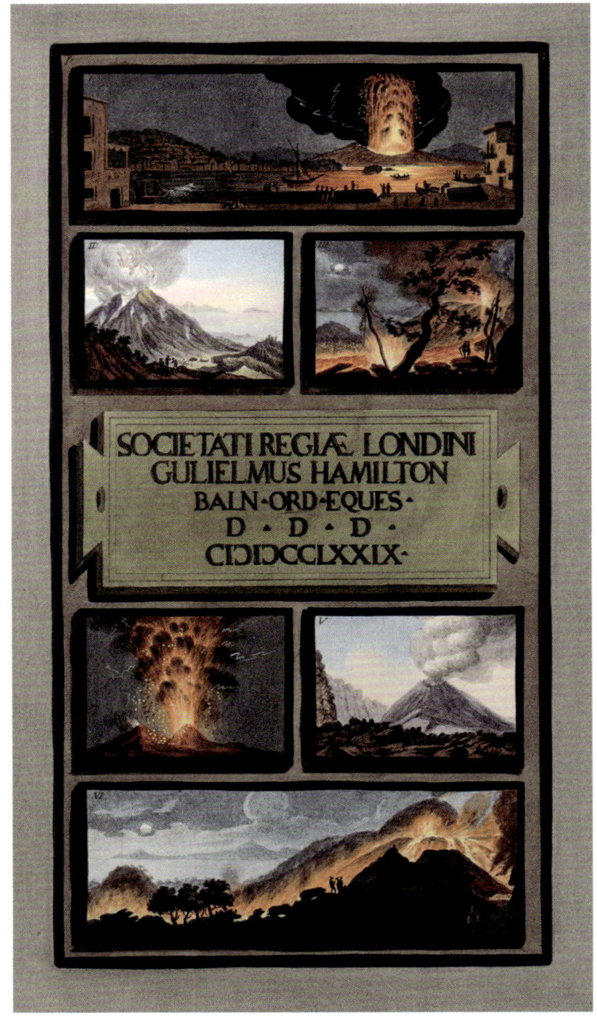

← ↑ *Campi phlegraei* (Flaming fields), 1776. William Hamilton (1730–1803), author; Pietro Fabris (active ca. 1740–1804), artist

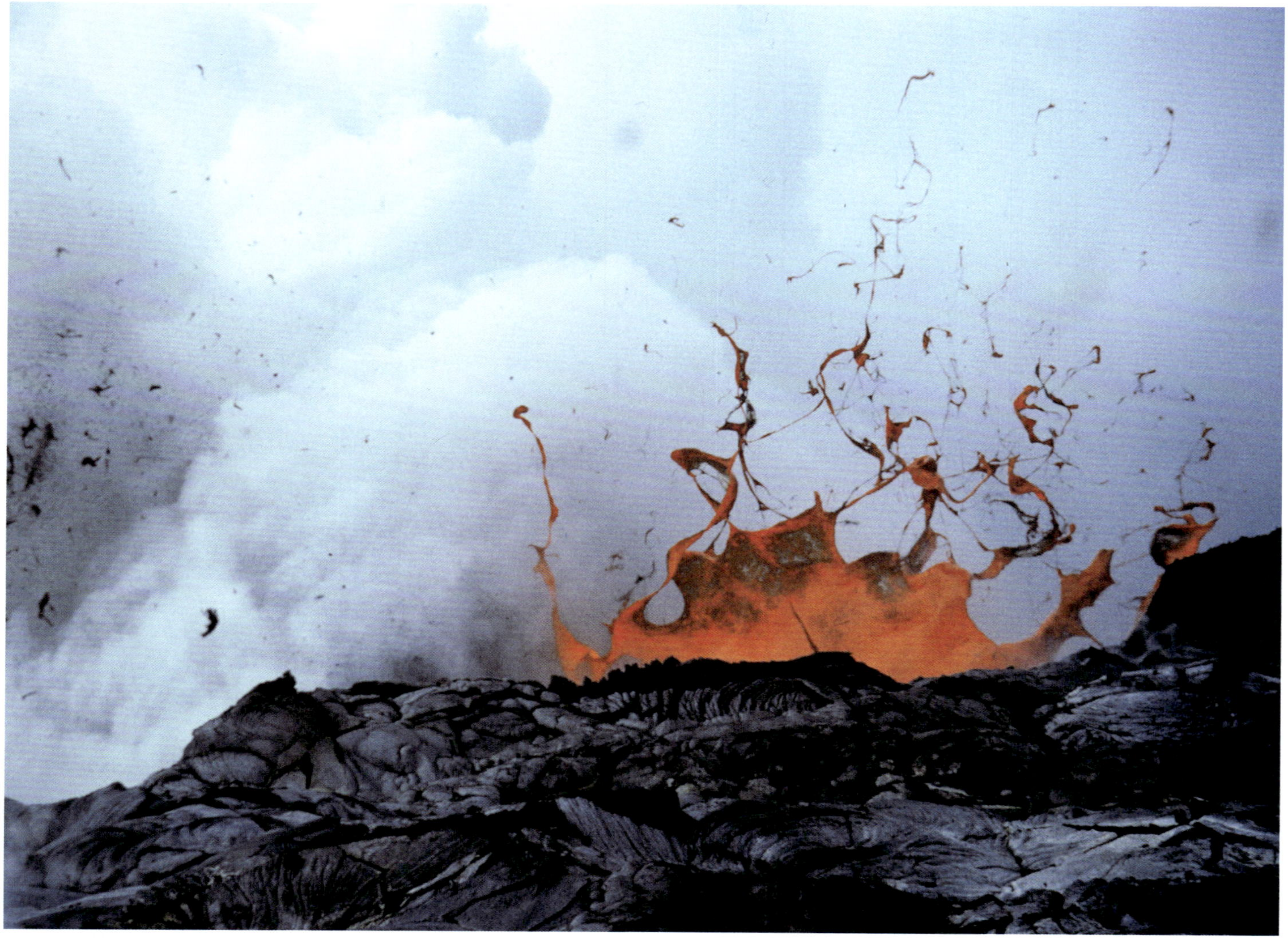

↑ *Lava bubbles at Kalapana, Hawaiʻi, 1988*

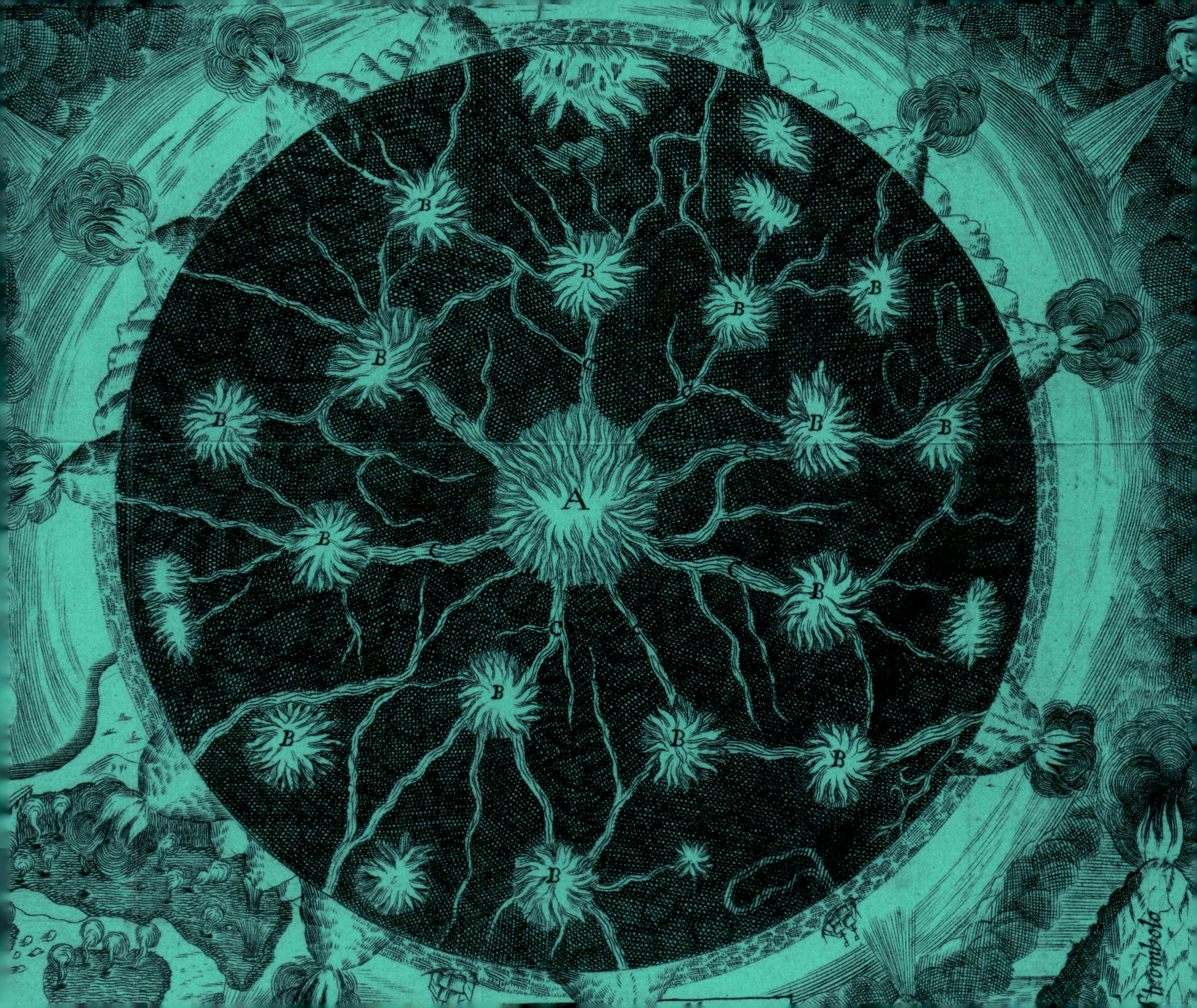
A
B
C

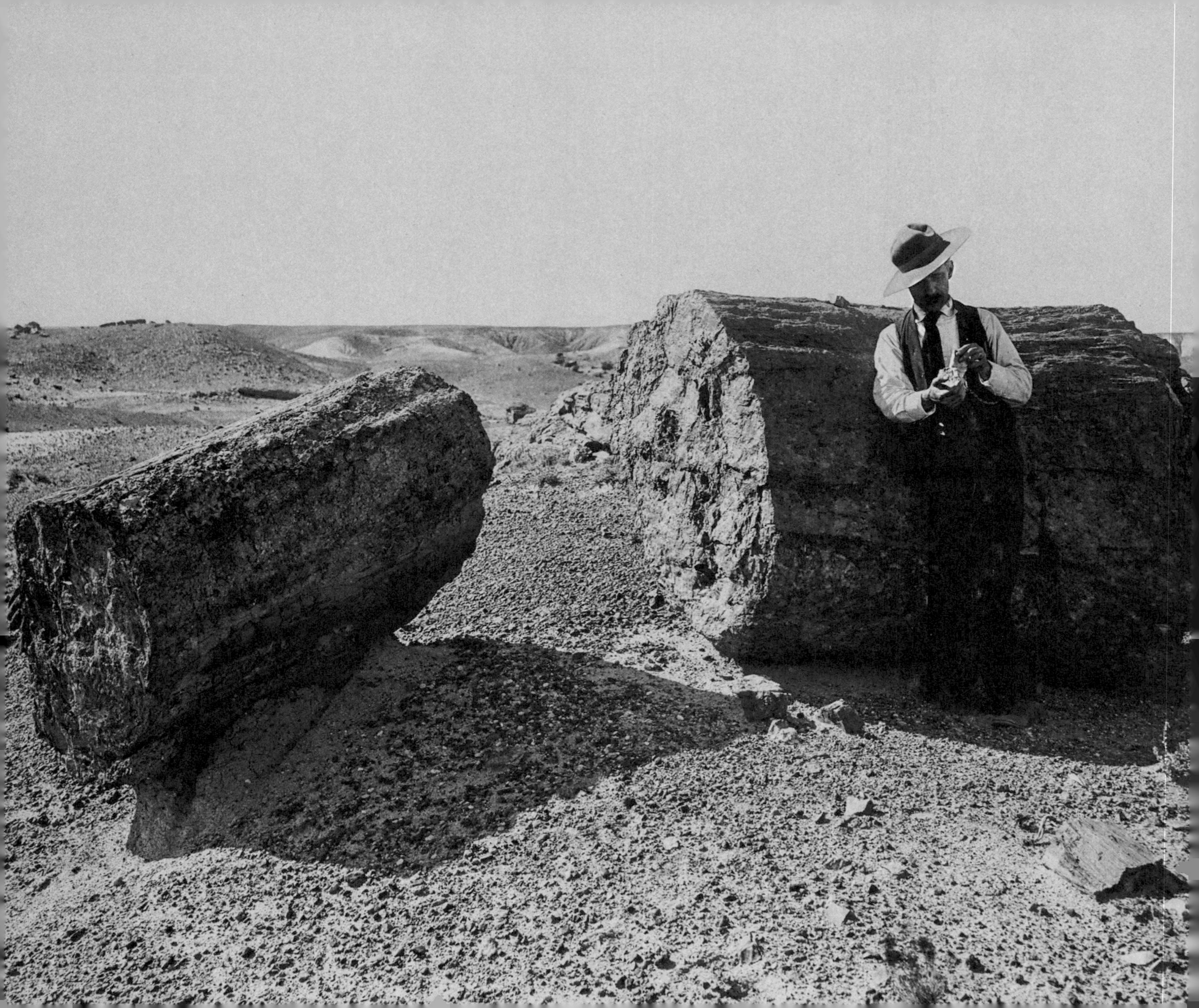

Dig

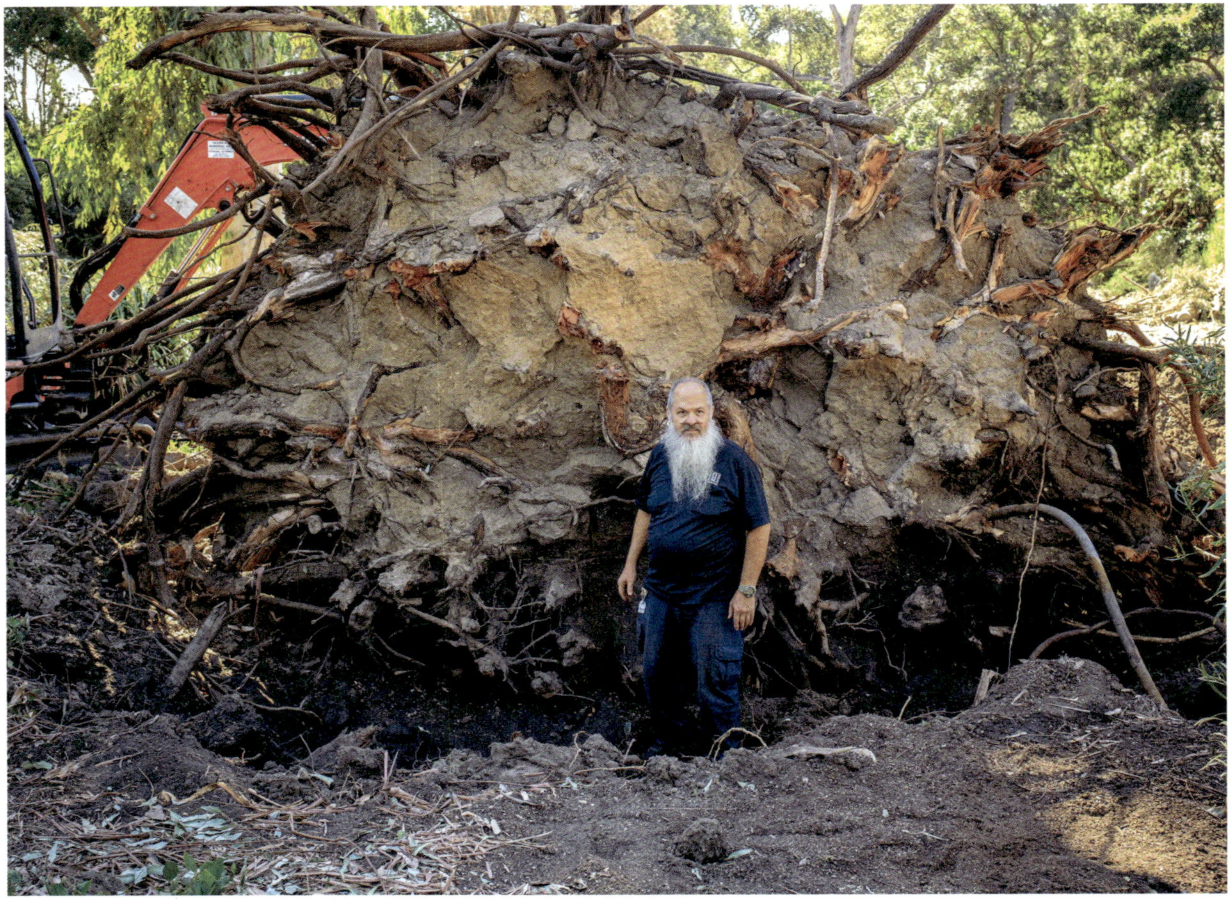

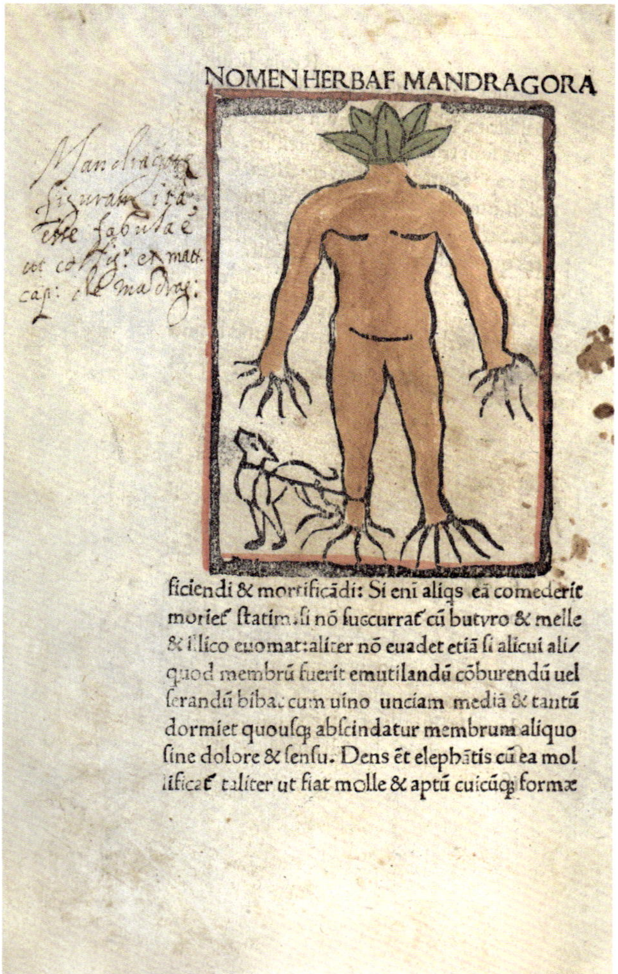
NOMEN HERBAE MANDRAGORA

ficiendi & mortificādi: Si eni aliqs eā comederit
morief statim.si nō succurraf cū butyro & melle
& illico euomat:aliter nō euadet etiā si alicui ali/
quod membrū fuerit emutilandū cōburendū uel
serandū bibat cum uino unciam mediā & tantū
dormiet quousq; abscindatur membrum aliquo
sine dolore & sensu. Dens ēt elephātis cū ea mol
lificaf taliter ut fiat molle & aptū cuicūq; formæ

Going Deep

⊙ 52
63

Hidden and often unappreciated, roots are essential to survival. A 15th-century herbal (book of plants) depicts a humanoid mandrake being pulled by a little dog. Early folklore said that, when uprooted, the mandrake emitted a scream capable of killing those within earshot. The solution? Sacrificing a canine to perform the deadly deed. Bonaventura Corti, an Italian priest and naturalist, captured roots at a microscopic level, revealing their delicate inner structures. The Huntington's Cryobiotechnology Lab preserves a variety of endangered plants by culturing shoot tips (apical meristems), cooling them in liquid nitrogen, and storing them long-term to safeguard genetic diversity and enable future regeneration.

↩ *Eucalyptus tree downed during Eaton Fire windstorm, with employee Damon Miears, The Huntington*, January 2025. Ibarionex Perello (born 1965)

← *Herbarium Apulei* (Herbarium of Apuleius), ca. 1481–82. Barbarus Apuleius (active 9th century)

→ *Osservazioni microscopiche sulla tremella* (Microscopic observations on tremella), 1774. Bonaventura Corti (1729–1813)

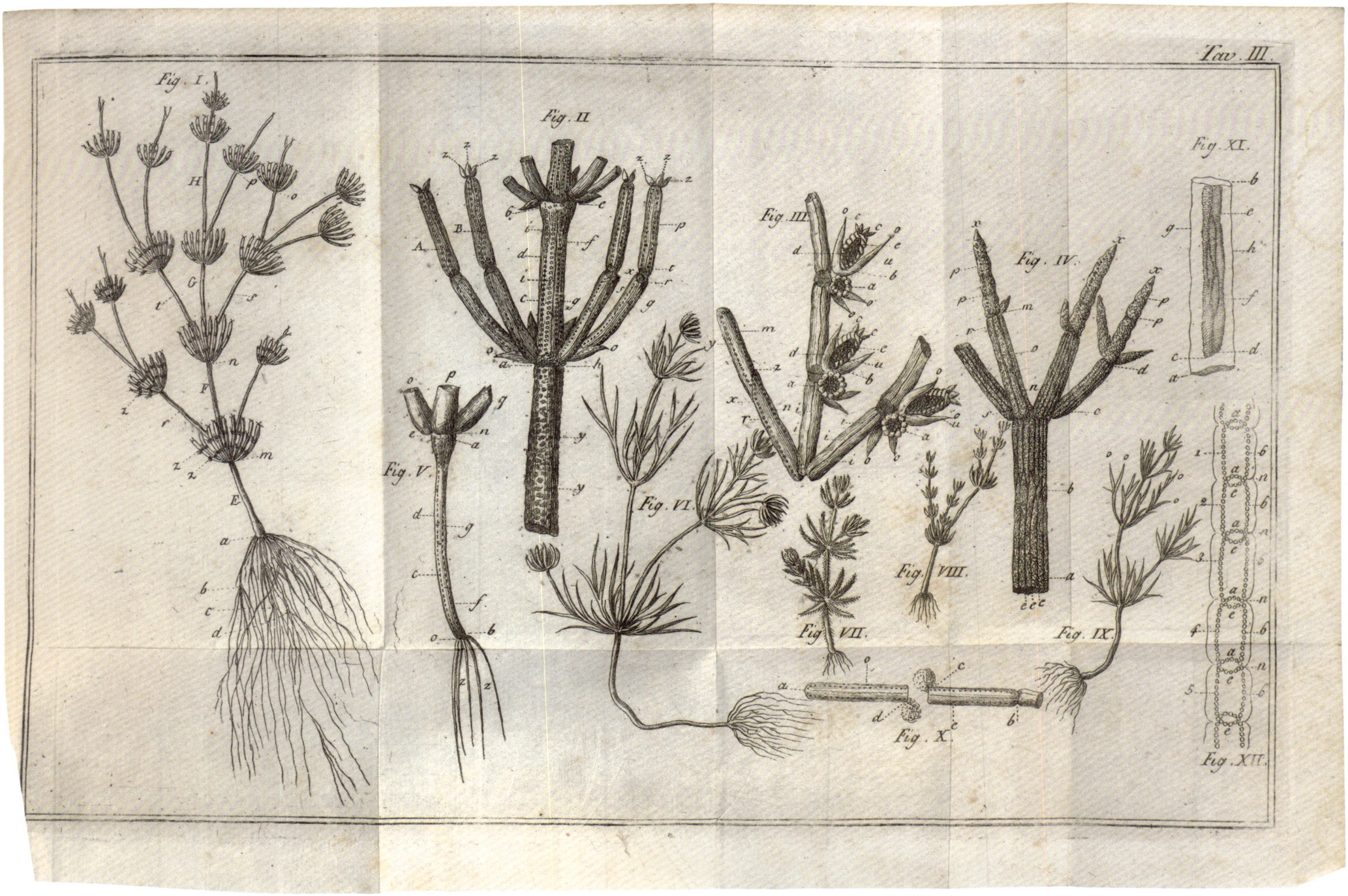
Tav. III.
Fig. I.
Fig. II
Fig. III
Fig. IV.
Fig. V.
Fig. VI.
Fig. VII.
Fig. VIII.
Fig. IX.
Fig. X.
Fig. XI.
Fig. XII.

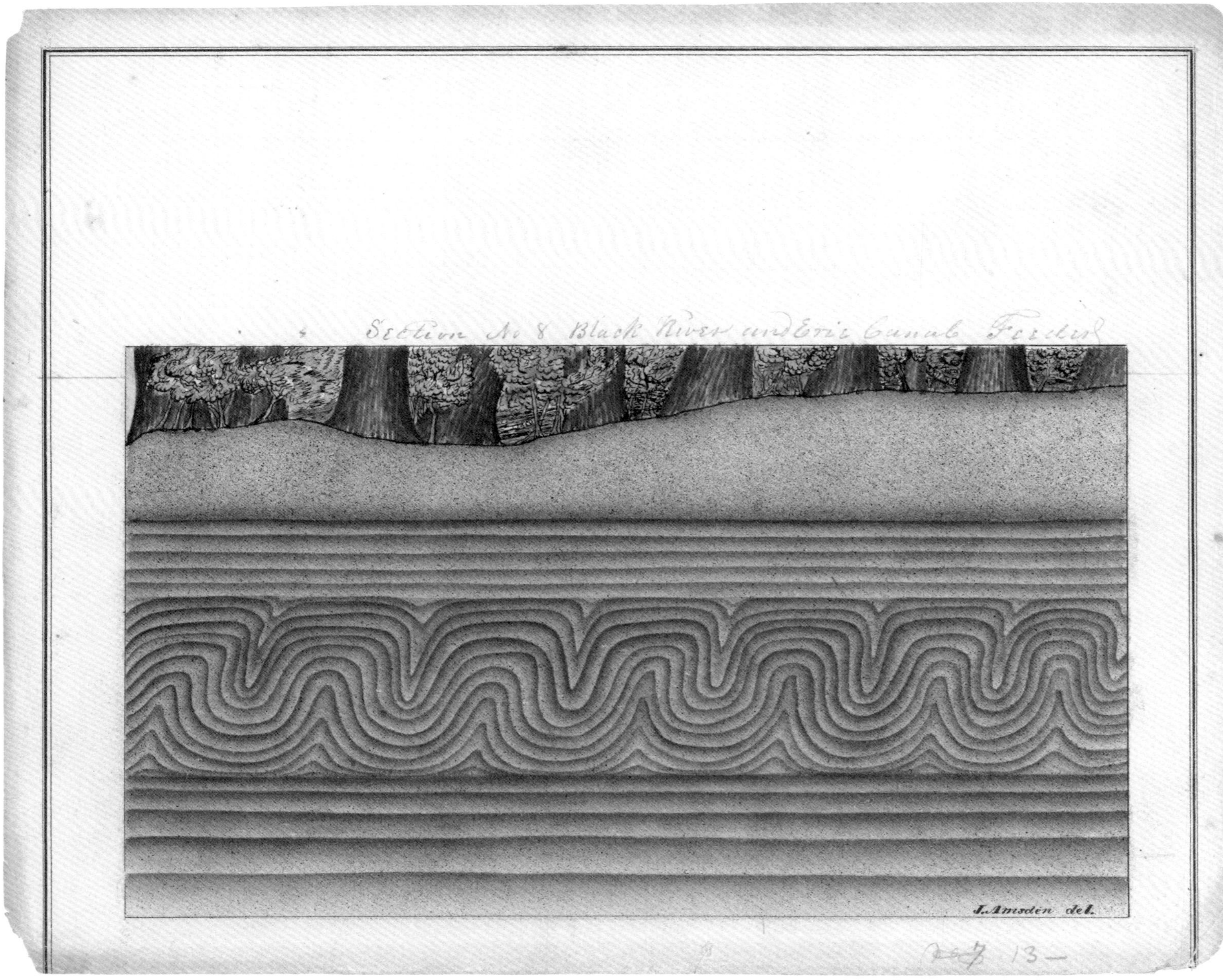
Section No 8 Black River and Erie Canal Feeder
J. Amsden del.
13—

A Matter of Time

Although mountains, valleys, and ocean floors seem static to the human eye, their compositions reflect vast transformations over time. In the late 18th and early 19th centuries studies across continents proved that Earth's geological processes are slow and continuous, implying immense age. Today, we know that layer upon layer of sedimentary rock speaks to 4.5 billion years of planetary evolution: tectonic shifts, volcanic disruptions, meteor impacts, and climate change.

FRONT FLAP

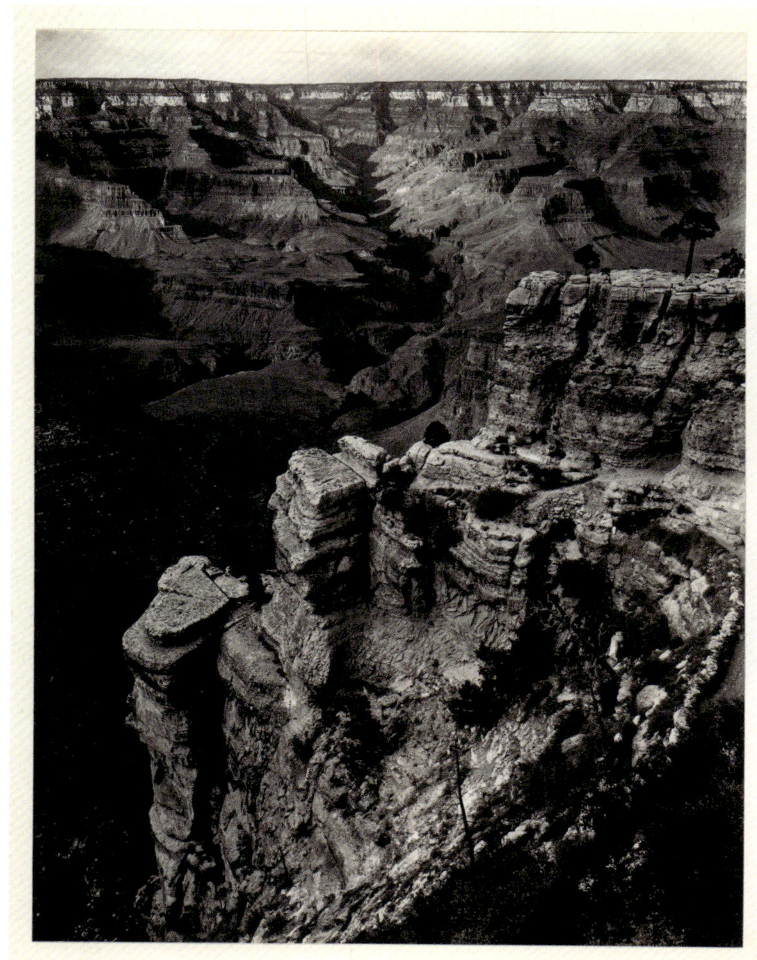

← Original sketch of *Section No. 8, Black River and Erie Canal Feeder, New York*, ca. 1841. Lardner Vanuxem (1792–1848)

→ *Grand Canyon*, ca. 1941. Edward Weston (1886–1958)

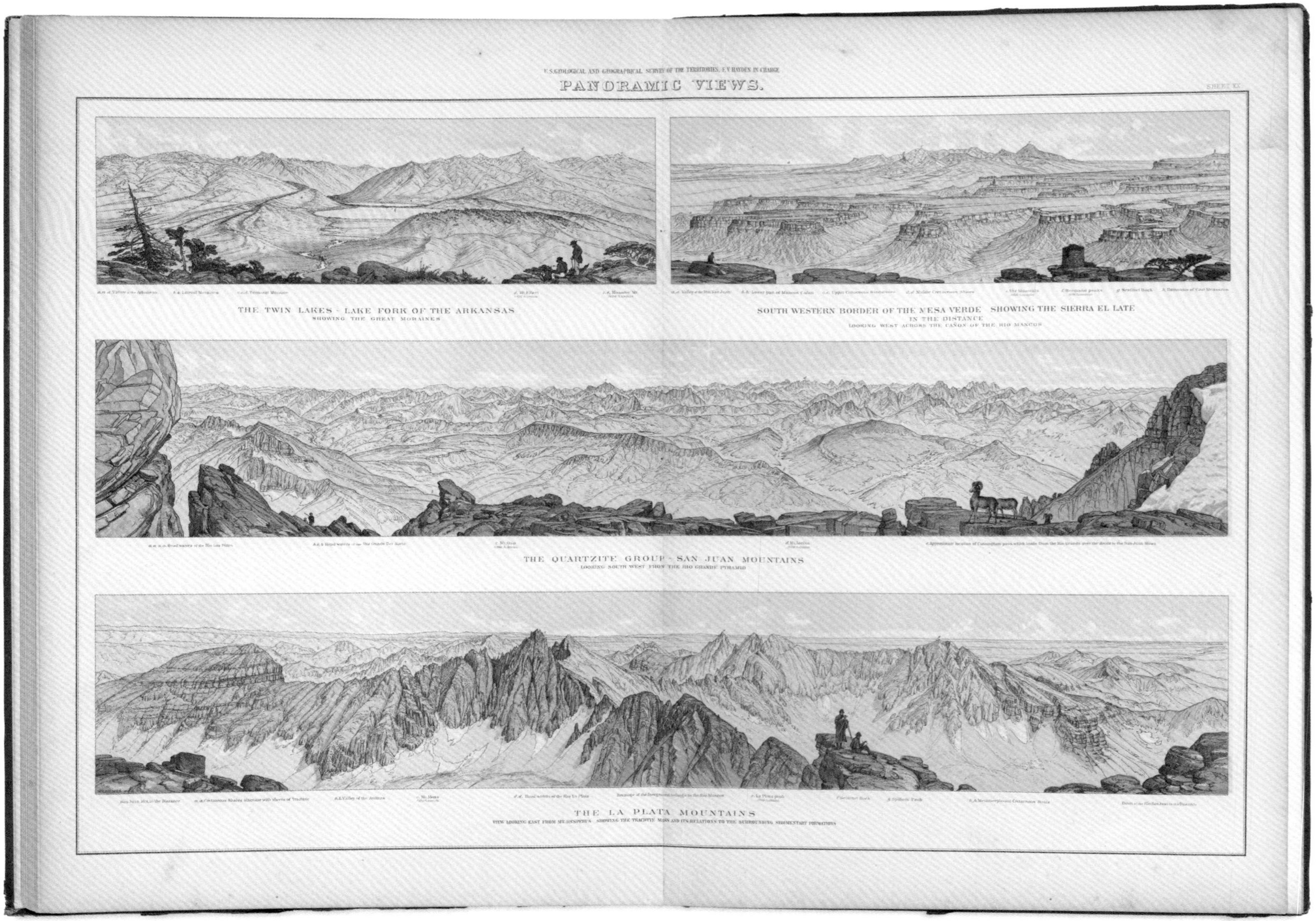

↑ *Geological and Geographical Atlas of Colorado and Portions of Adjacent Territory*, 1877. F. V. Hayden (1829–1887)

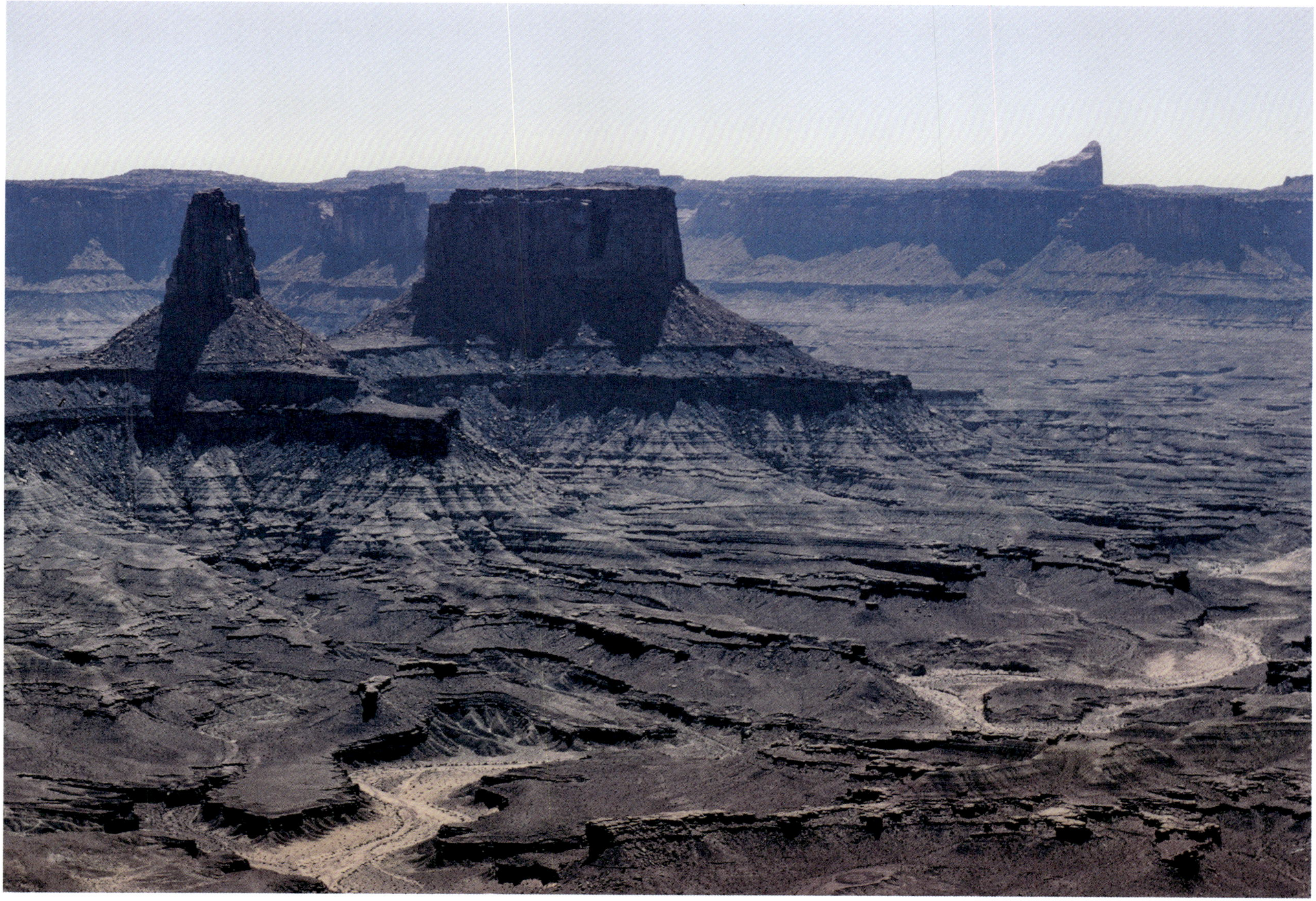

↑ *Buttes of the Cross, Cleopatra's Chair, Utah*, 1965.
Otis R. Marston (1894–1979)

Petrified Wood

Fossilized trees are scattered across 147 miles of northeastern Arizona, colorful remnants of the region's ancient tropical past. By the late 19th century, these rainbow-hued rocks—wood pressurized over millennia under sediment and ash—lured tourists, jewelers, and gem seekers, who often pocketed what they found. Despite the area becoming a protected monument in 1906 and later Petrified Forest National Park, visitors continued to have itchy fingers. Pangs of guilt, remorse, and fear compel some of these pilferers to send their petrified mementos back.

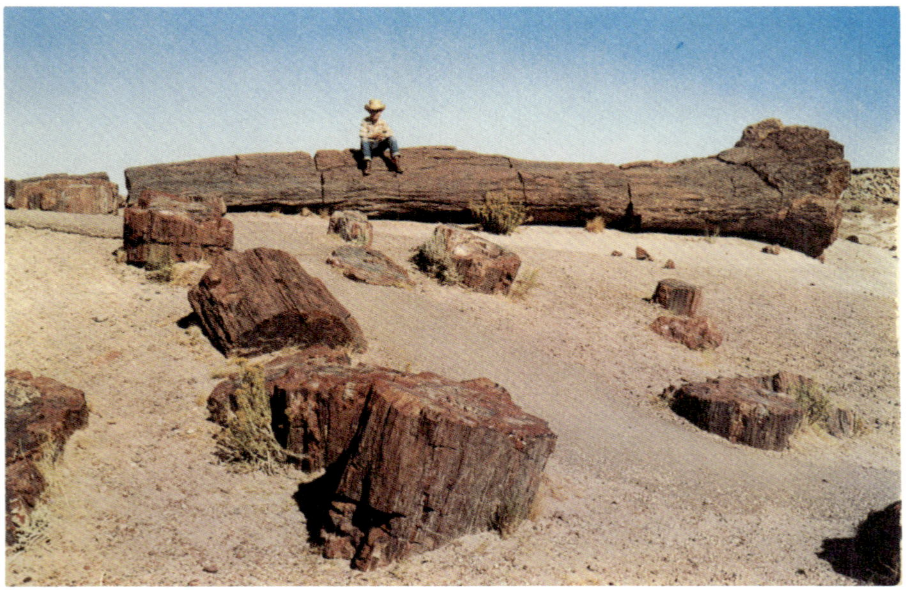

← Postcard of Petrified Forest, near Holbrook Arizona, ca. 1950. Bob Bradshaw (1918–2008)

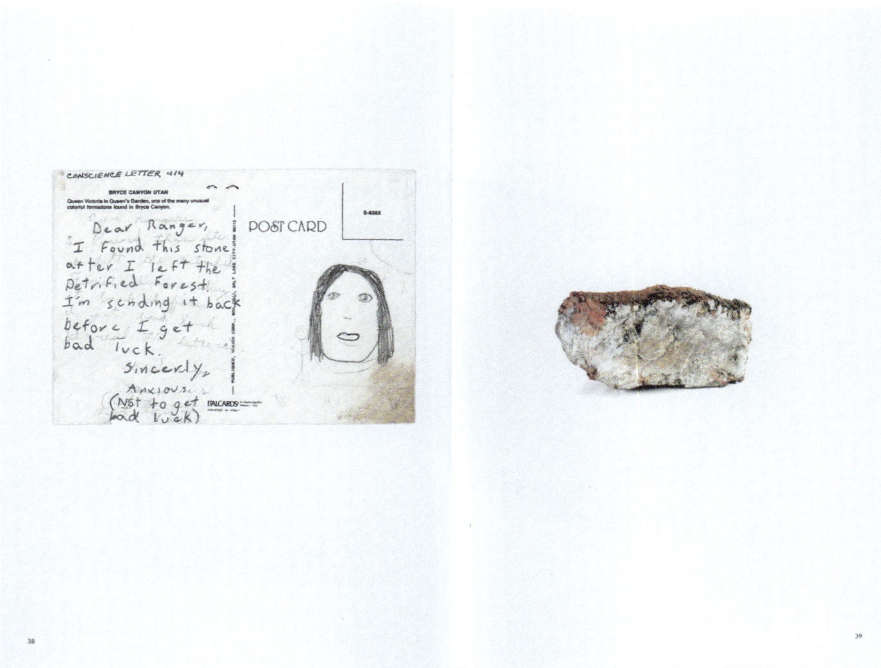

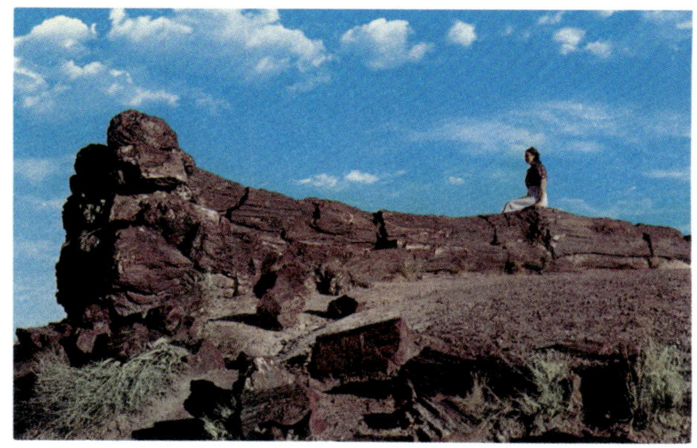

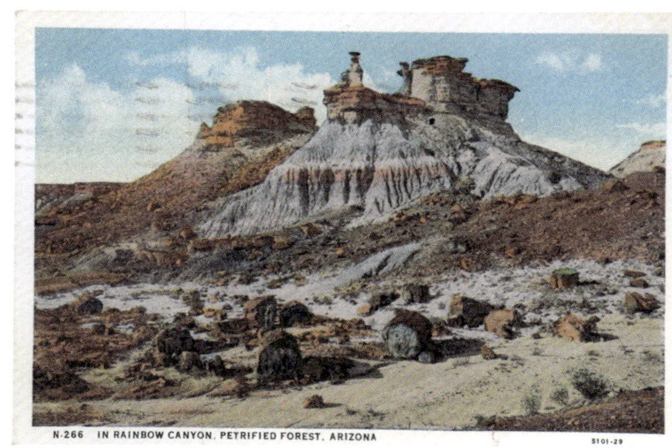

↑ *Bad Luck, Hot Rocks*, 2014. Ryan Thompson (born 1981) and Phil Orr

→ (TOP) Postcard of Old Faithful Log, Petrified Forest, Arizona, ca. 1945; (BOTTOM) Postcard of Rainbow Canyon, Petrified Forest, Arizona, ca. 1933. Curt Teich & Co.

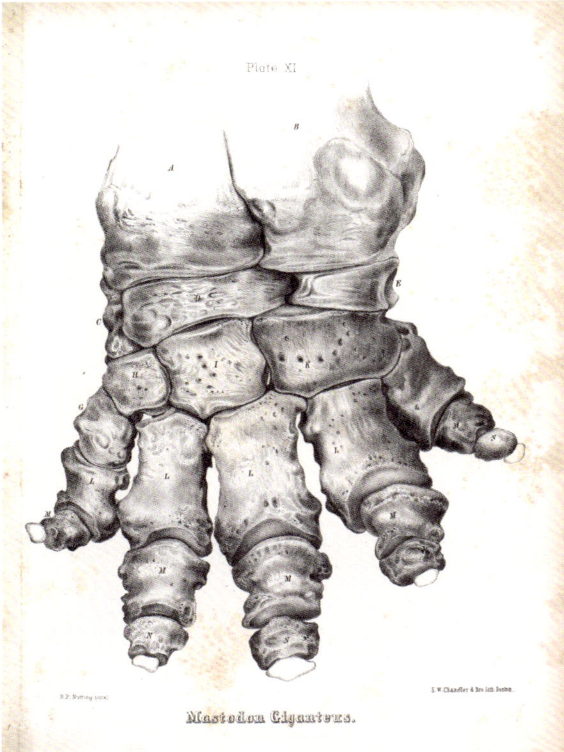

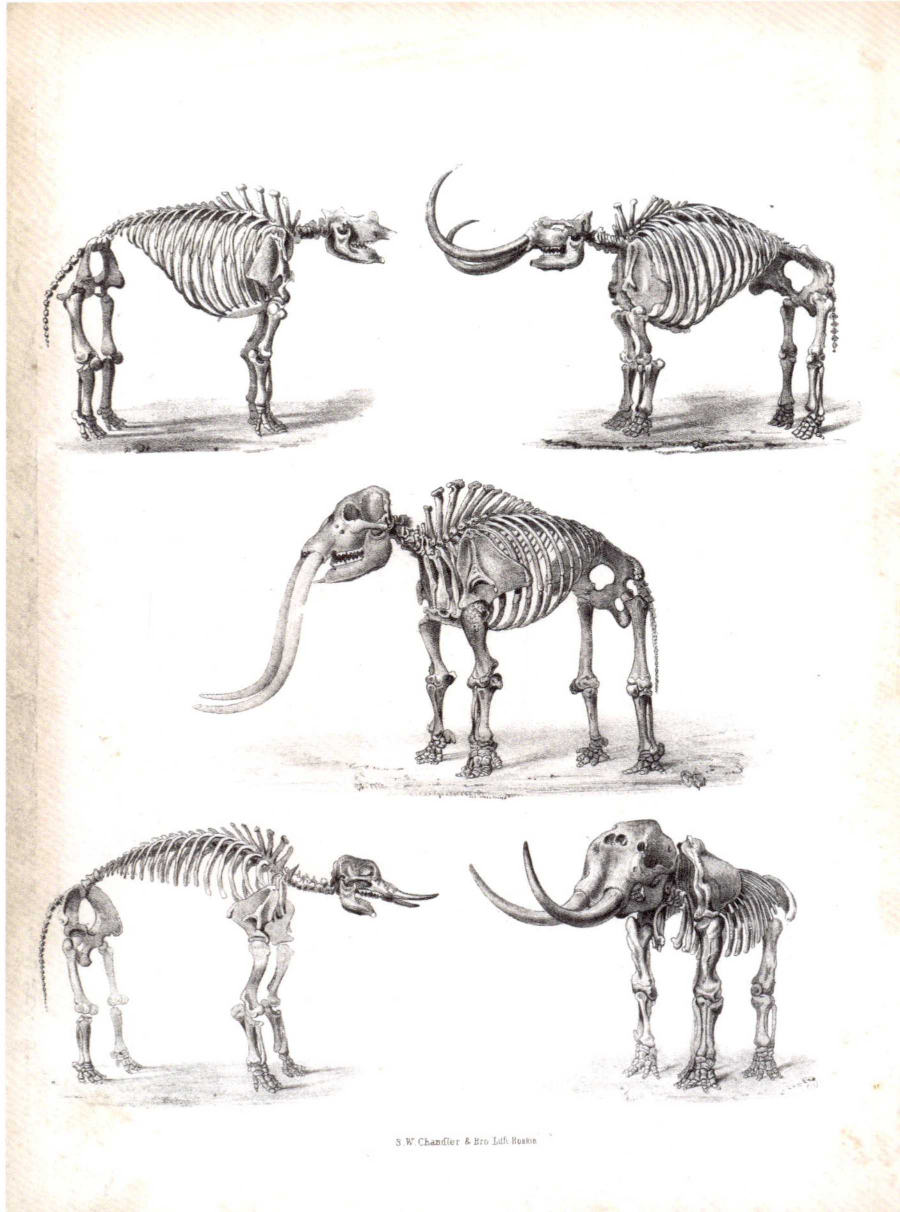

↑ → *Description of a Skeleton of the Mastodon Giganteus of North America*, 1852. John Warren (1778–1856), author; Benjamin Franklin Nutting (ca. 1826–1884), artist

In 1845, the physician and anatomy professor John Warren rushed to an archaeological site in Newburgh, New York, to see the exposed bones of a colossal creature at the bottom of a dry lake. It was the most complete skeleton of a mastodon ever documented. Warren purchased the bones, hoping to preserve and display them in his own museum of wonders. Today, the "Warren mastodon" stands at the American Museum of Natural History, where it has towered over visitors for more than a century.

50–52

Fossil Record

While exploring South America aboard the HMS *Beagle*, Charles Darwin unearthed the fossilized bones of enormous extinct animals, including a strange long-necked creature he described as "a gigantic guanaco"—a prehistoric relative of the camel and llama. These remains hinted at vanished worlds and the slow transformation of species over time. Nearly two centuries earlier, the anatomist Nicolaus Steno drew a similar conclusion from the jaw of a giant shark. Upon dissecting the fish, he recognized that its teeth matched the so-called tongue stones found in rock, once believed to be the petrified tongues of serpents or charms that fell from the sky during eclipses.

⊙ 68

50

→ *The Zoology of the Voyage of H.M.S.* Beagle*, Part I: Fossil Mammalia*, 1839–43. Charles Darwin (1809–1882), editor; Richard Owen (1804–1892), author; George Johann Scharf (1788–1860), artist

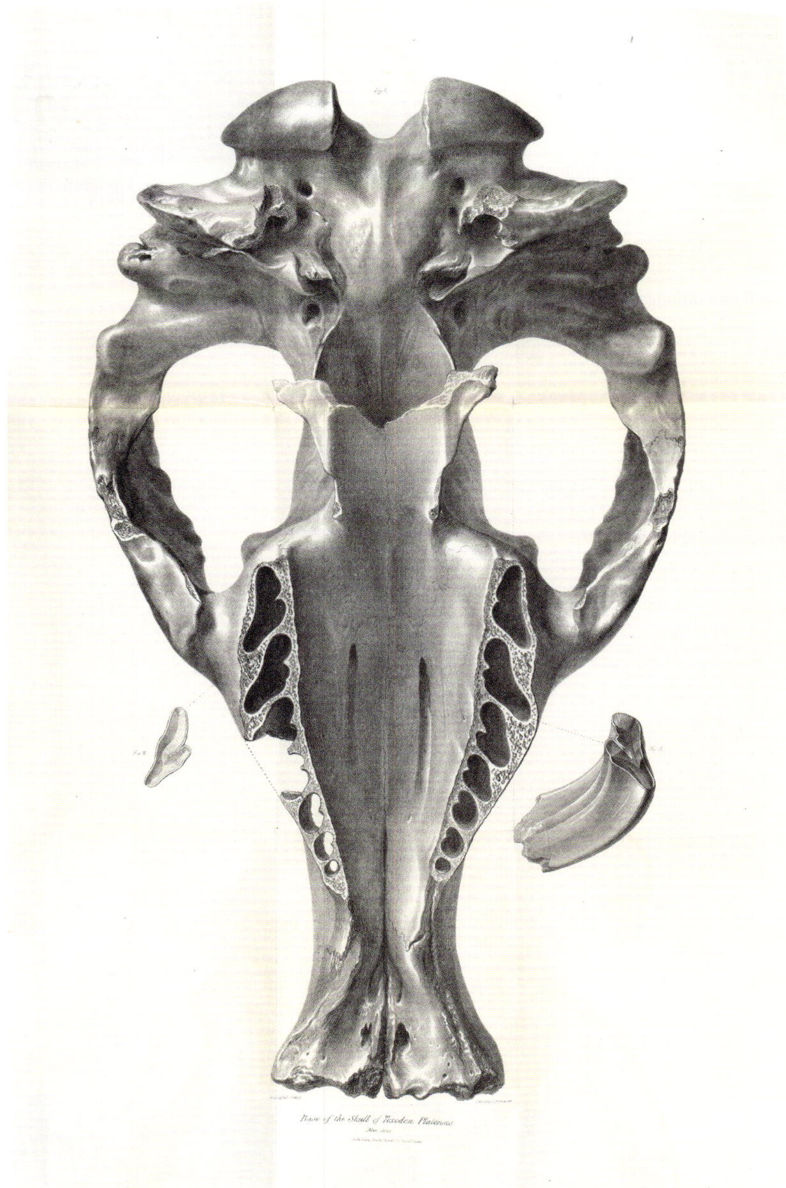

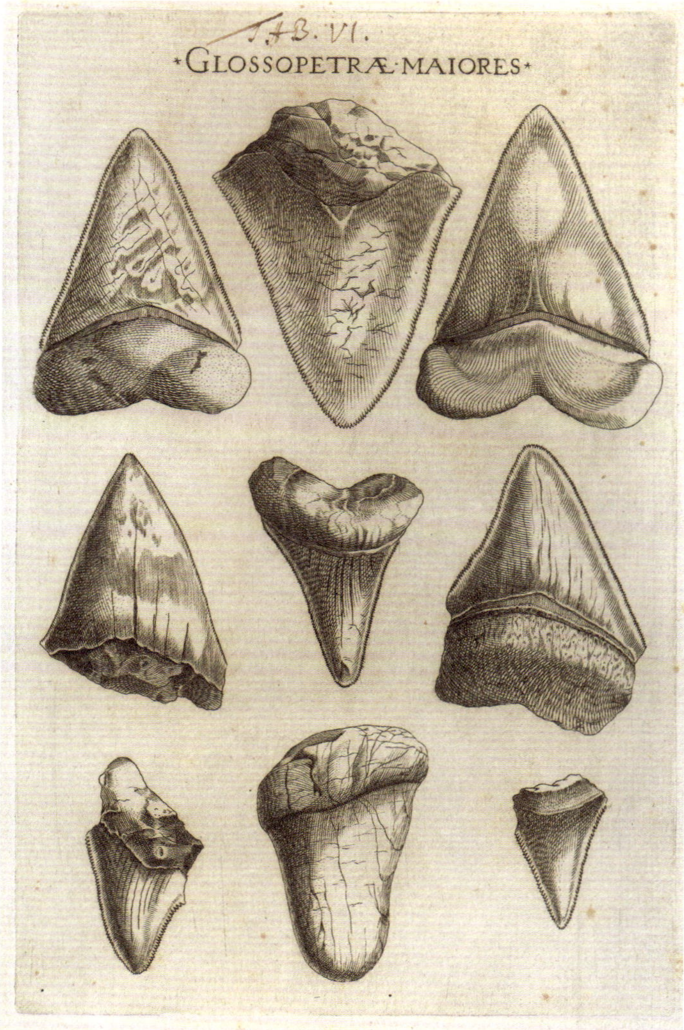

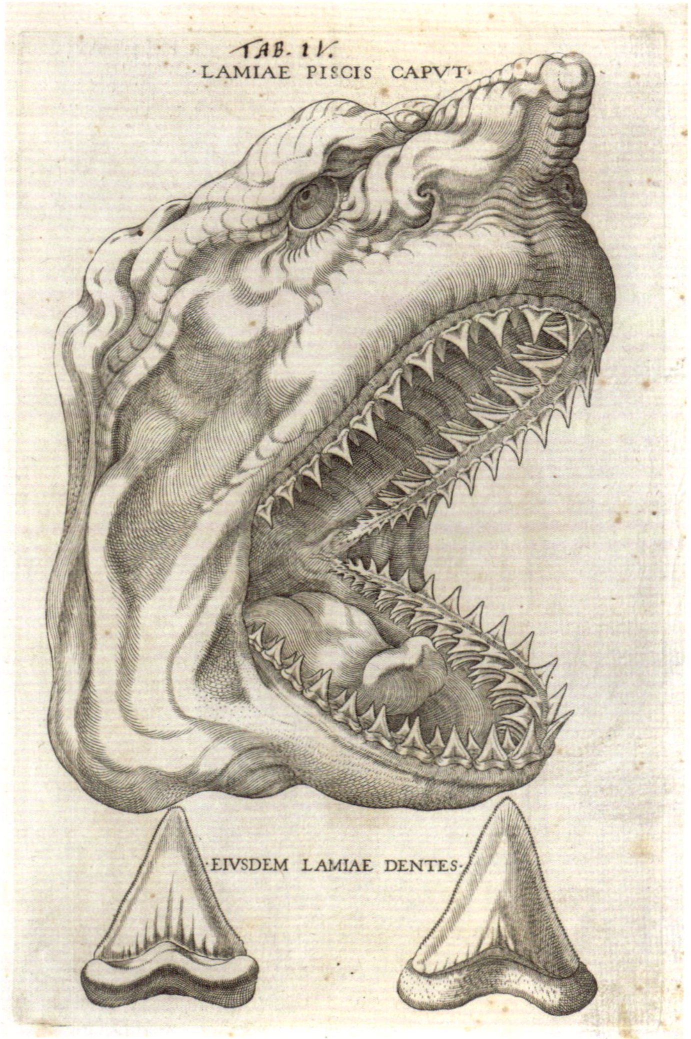

↑ *Nicolai stenonis elementorum myologiae specimen* (Specimen of the elements of myology), 1667. Nicolaus Steno (1638–1686)

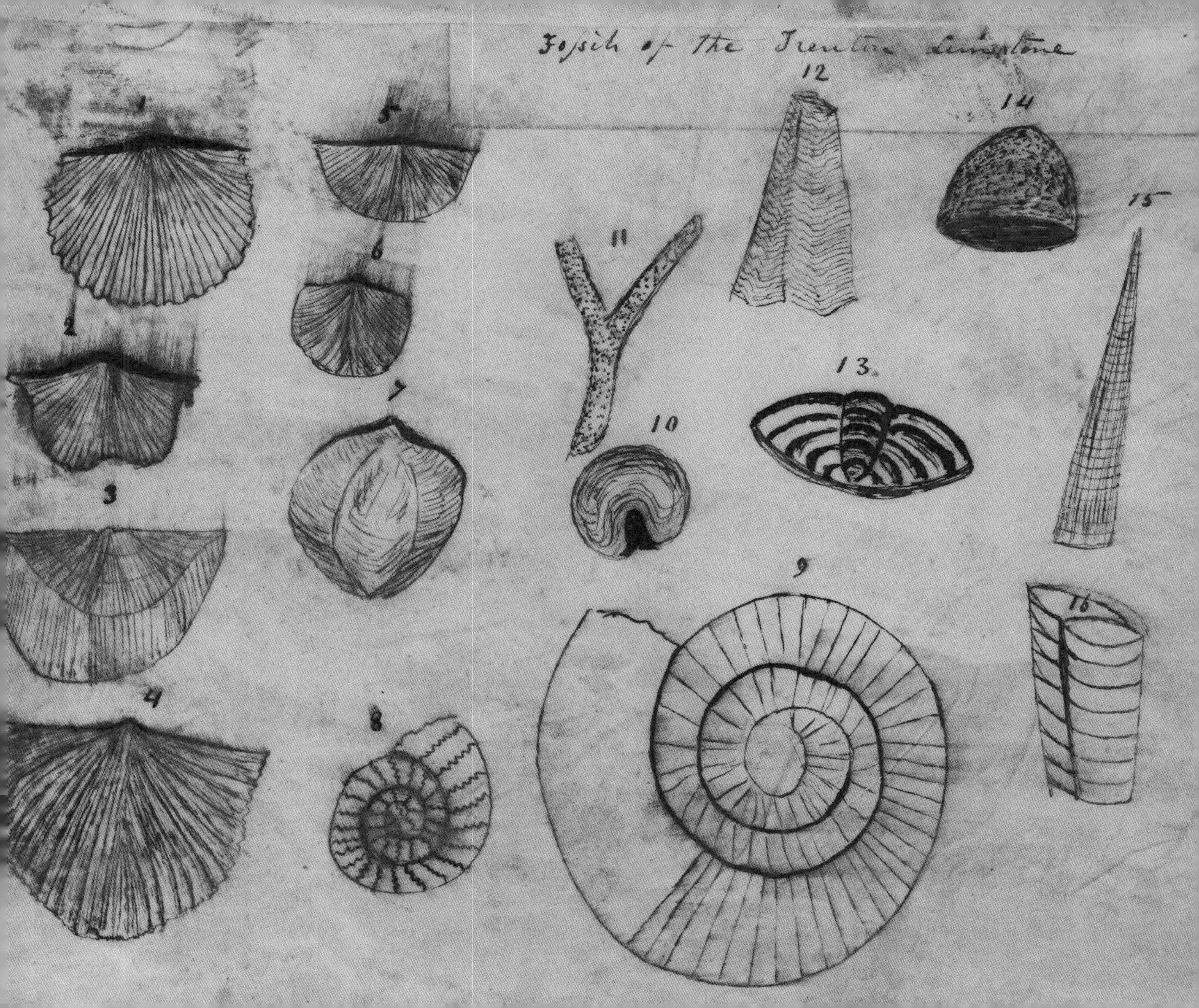
Fossils of the Trenton Limestone
1
2
3
4
5
6
7
8
9
10
11
12
13.
14
15
16

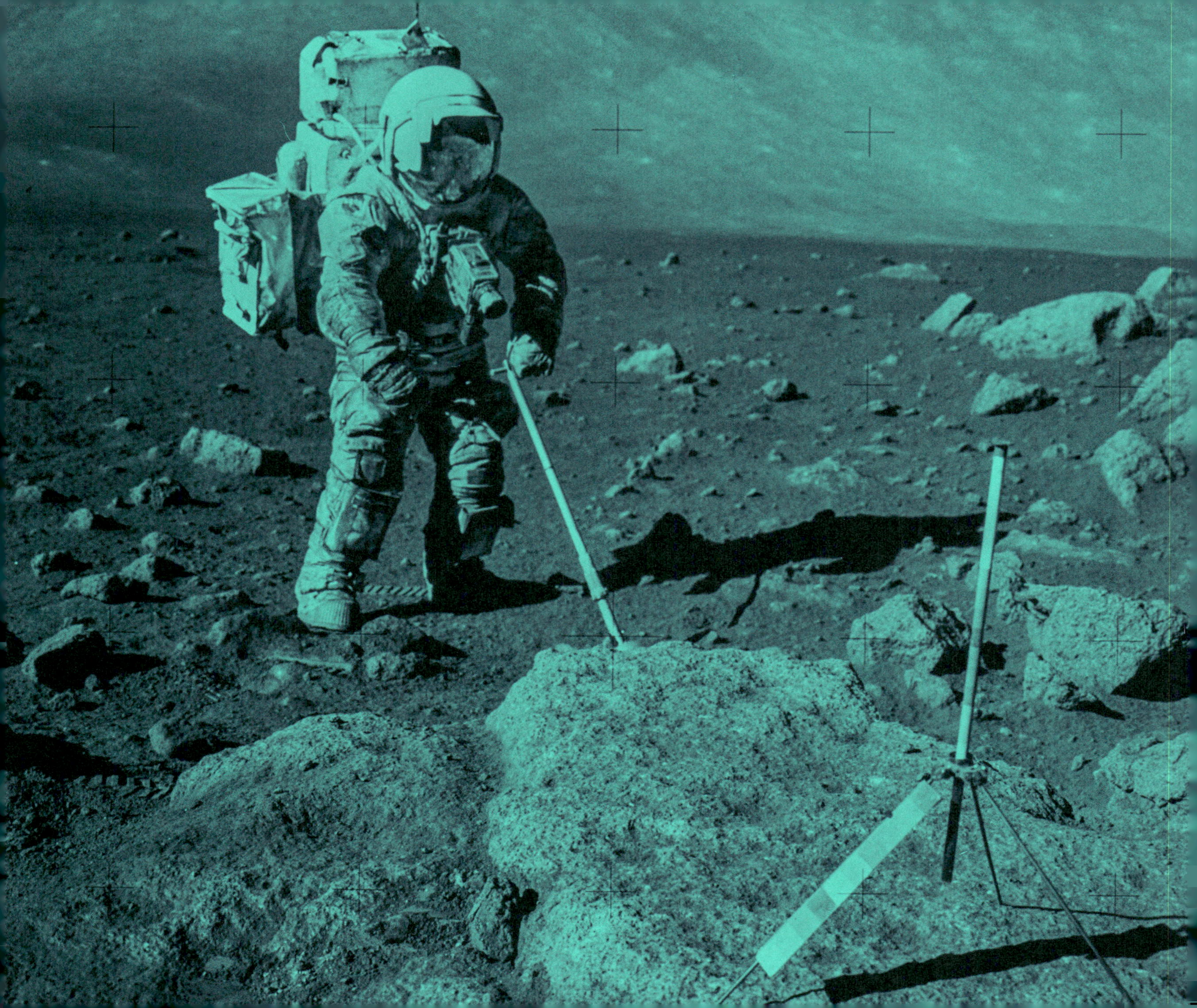

Survey

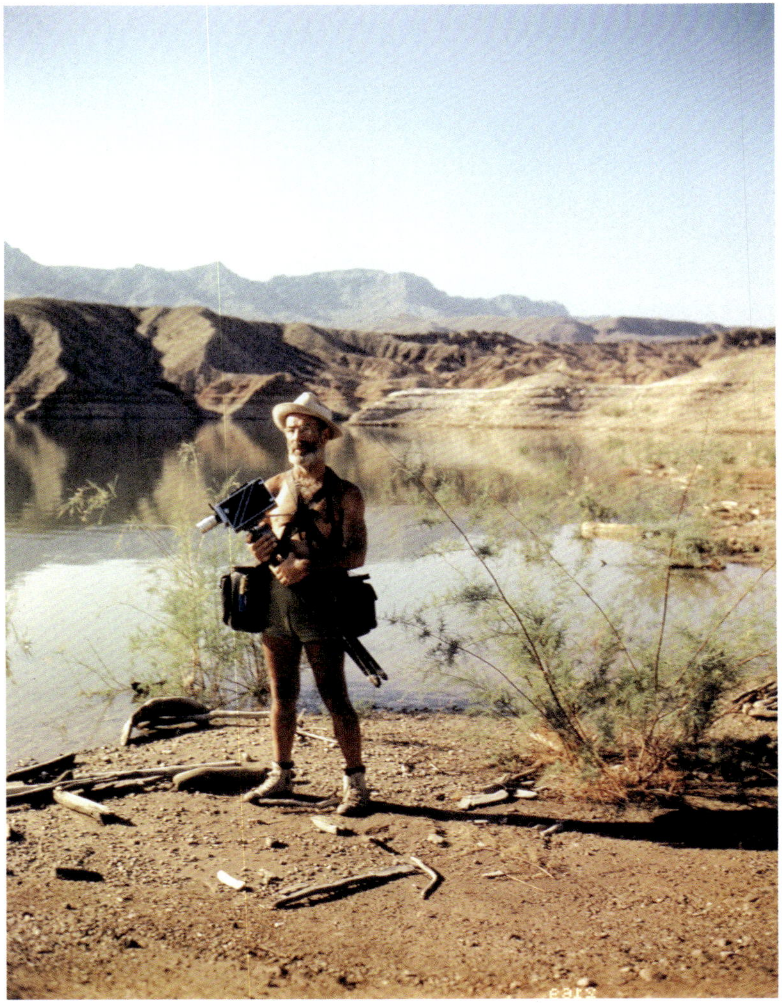

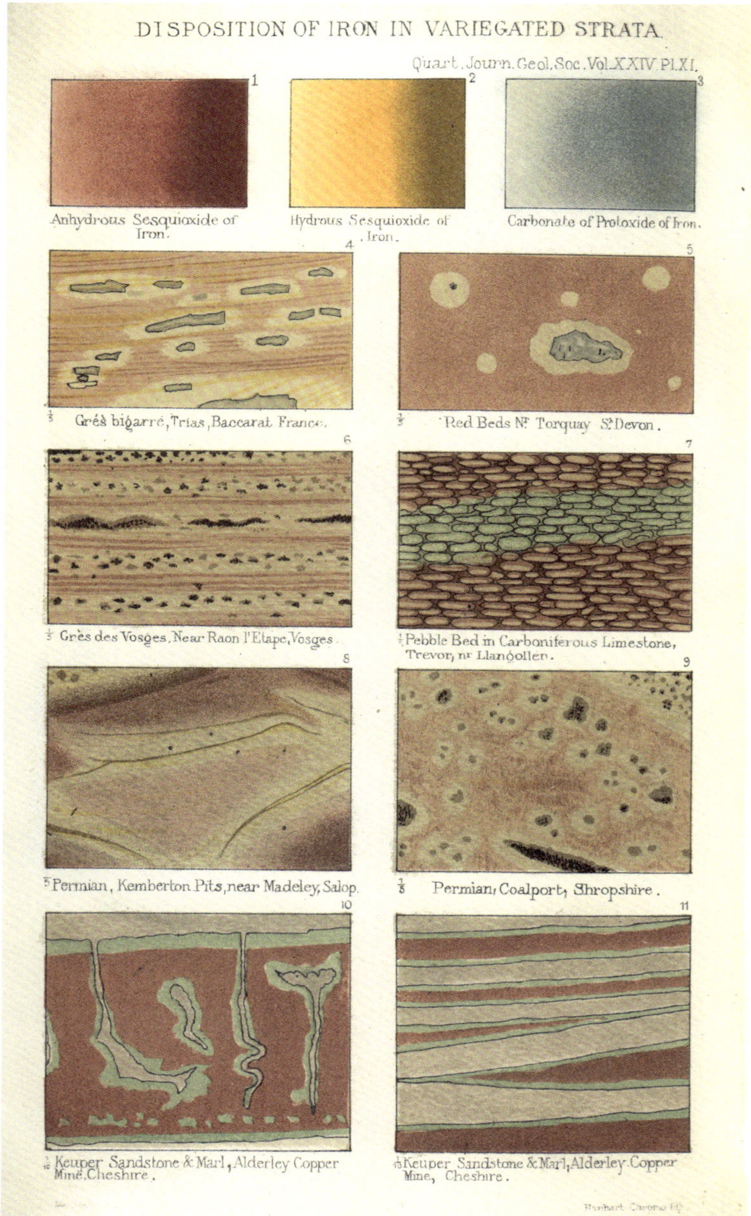

Granular Details

Under a microscope, small geologic samples become worlds unto themselves, revealing intricate contours, coloring, and mineral composition. Beginning in the 1860s, European ⩡ 26–29 geologists honed microscopic analysis to accurately record ≈ 4 and classify rock samples, adding granular depth to Earth's layers or stratigraphy. This was fastidious work. Samples needed to be polished flat or planed into thin sections measuring nearly a thousandth of an inch, permitting light to transmit through. To draw what they observed, the geologist's eye darted quickly between the instrument's eyepiece and a sketchpad close at hand.

↩ *Colorado River runner Otis "Dock" Marston at Pearce Ferry, Grand Canyon,* 1947. Margaret Marston (1892–1968)

← *On the Disposition of Iron in Variegated Strata*, 1868. George Maw (1832–1912)

→ *Minéralogie micrographique: Roches éruptives françaises* (Micrographic mineralogy: French igneous rocks), 1879. Ferdinand Fouqué (1828–1904) and Auguste Michel Lévy (1844–1911), authors; M. Jacquemin, artist

PLANCHE XL

FIG. 1. — NODULE A OLIVINE DANS LE BASALTE D'ANLIAC, HAUTE-LOIRE

Gross. = 80 diam. Lumière polarisée, nicols croisés.

Cette roche constitue des coulées antérieures au creusement des vallées, postérieures aux andésites miocènes. C'est le basalte des plateaux. L'échantillon choisi provient d'une des dernières éruptions de ce basalte : il reste à Anliac des traces évidentes de cratère, des scories et des bombes.

Le dessin représente la matière du nodule et celle de la roche proprement dite : le nodule est composé d'olivine (23) ; d'enstatite (22) et de picotite (27) ; à la lumière naturelle la picotite est d'un brun verdâtre. La structure du nodule est granitoïde ; on n'y voit ni matière amorphe, ni microlithes ; il y a grande analogie avec les lherzolithes.

La roche proprement dite comprend les minéraux suivants :

I. Picotite (27) ; seulement dans le voisinage des nodules, fer oxydulé, olivine, augite (20).

II. Microlithes de fer oxydulé, d'augite, de labrador (7).

FIG. 2. — ANDÉSITE A MICA NOIR DE LA CASCADE DE TOURCI, CANTAL

Gross. = 80 diam. Lumière polarisée, nicols à 45°.

Cette roche se trouve en coulée entre l'andésite à sanidine (var. domite) et l'andésite à amphibole de la région. Elle paraît se rattacher à la partie supérieure de la domite.

Le dessin représente les minéraux suivants :

I. Pléonaste (27) ; mica noir (19) ; augite (20) ; labrador (7).

II. Microlithes de fer oxydulé, d'oligoclase (6) ; microlithes rares d'albite (5).

III. Vacuoles remplies de calcite (49), avec couronne d'opale (33).

Le pléonaste ne se rencontre dans cette roche que dans des agrégats de grands cristaux de labrador. Il est possible que ces agrégats aient été arrachés à une roche préexistante.

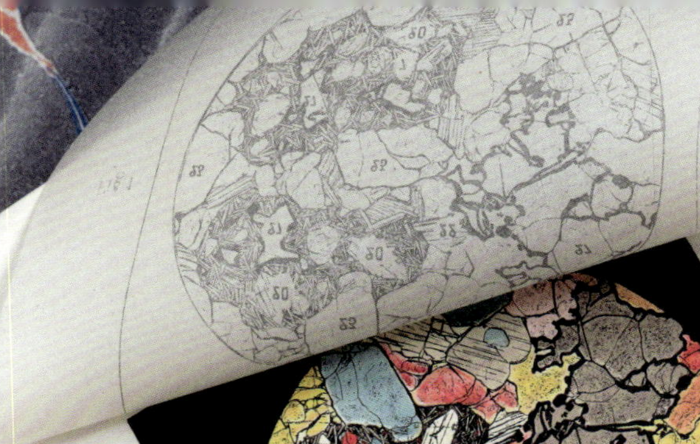

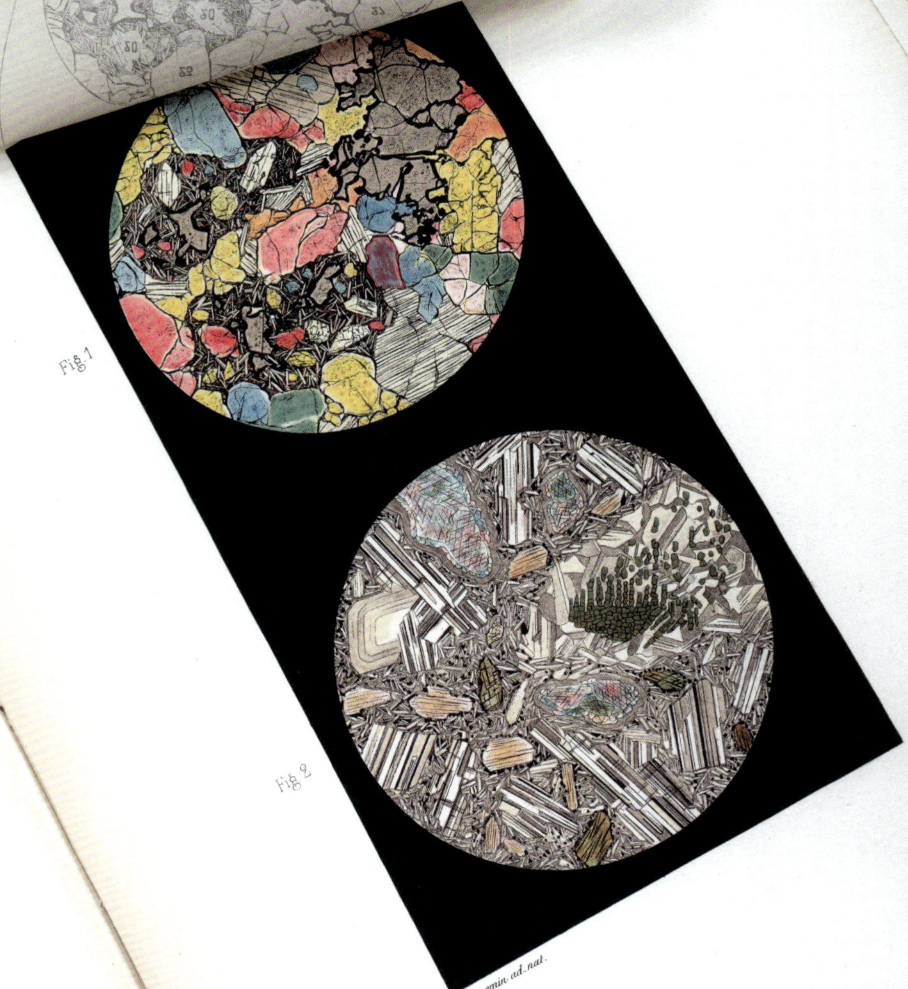

Fig. 1

Fig. 2

Jacquemin ad. nat.

Ateliers de Reproductions Artistiques

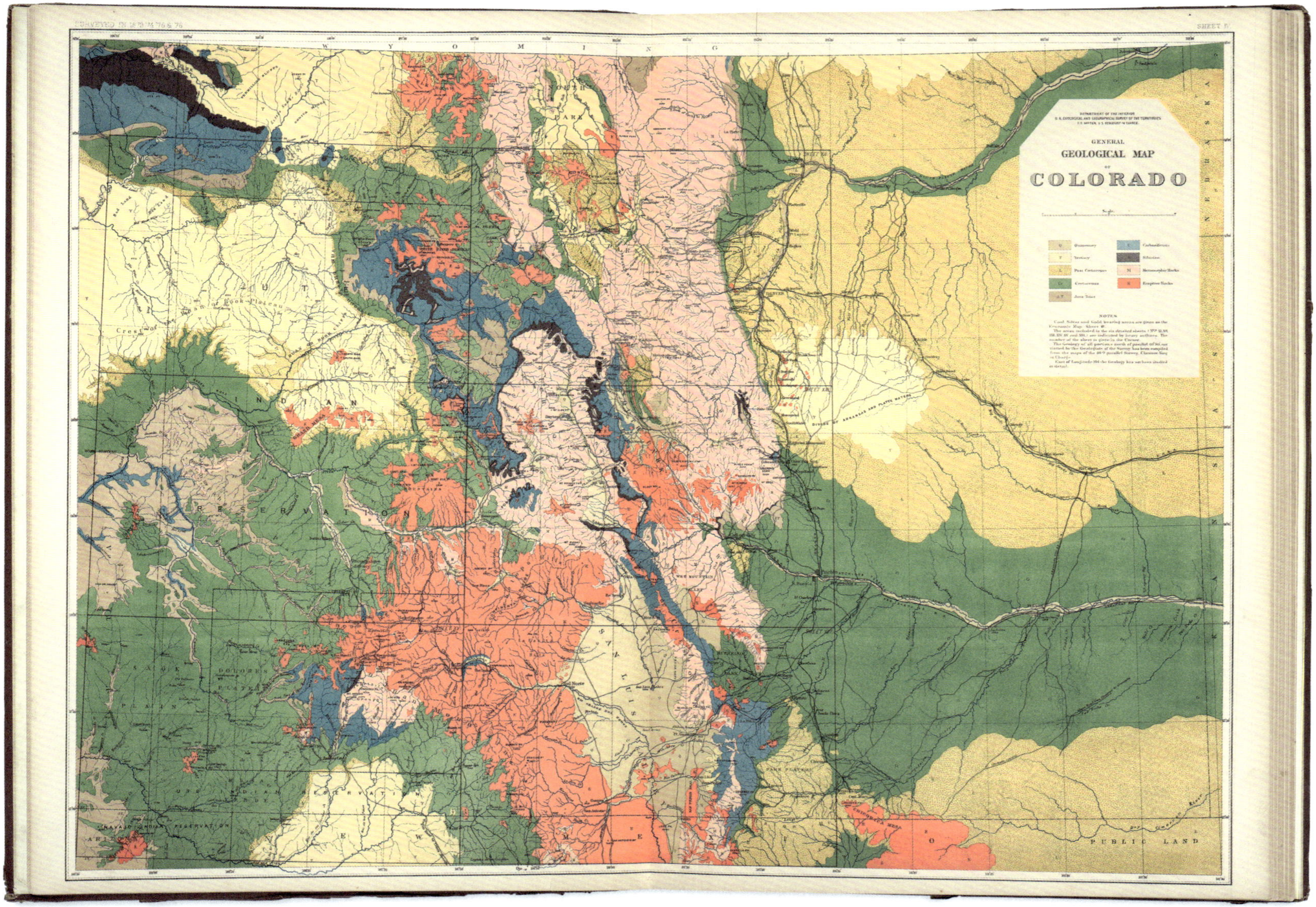
DEPARTMENT OF THE INTERIOR
U. S. GEOLOGICAL AND GEOGRAPHICAL SURVEY OF THE TERRITORIES
F. V. HAYDEN, U. S. GEOLOGIST IN CHARGE.
GENERAL
GEOLOGICAL MAP
OF
COLORADO
Scale.
NOTES
W Y O M I N G
N E B R A S K A
R E S E R V A T I O N
P U B L I C L A N D
SHEET

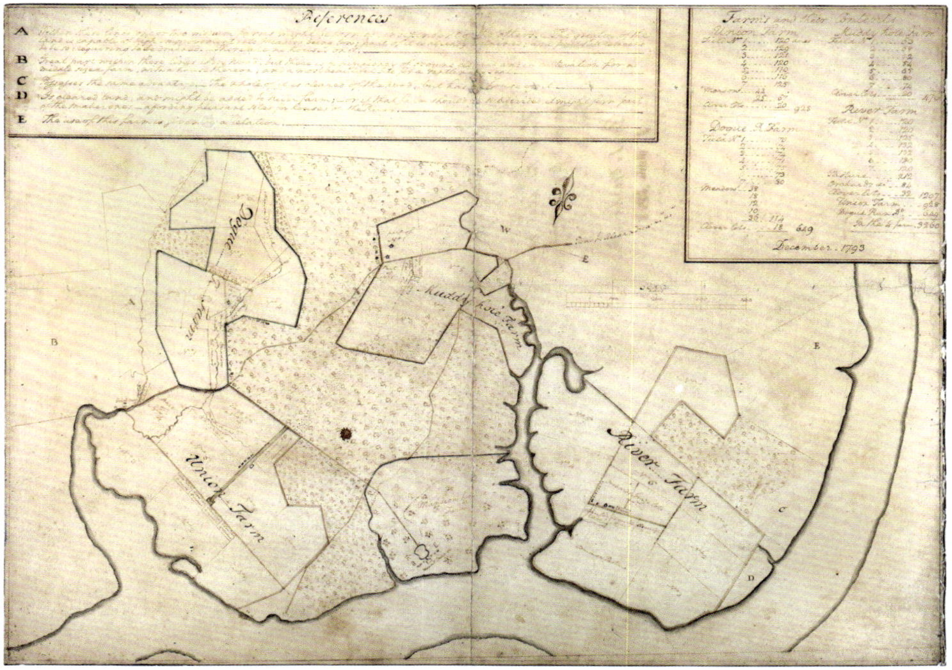

← Survey and plot of Mount Vernon and neighboring farms, 1793. George Washington (1732–1799)

Surveying is not just taking measurements and making maps: It defines power and place, marking out boundaries that stake claims and solidify ownership. George Washington began surveying as a teenager in Virginia's frontier, inscribing borders onto contested Indigenous land. This skill later facilitated his canny purchase of vast properties. While he was president, Washington surveyed the five farms and adjacent plots comprising Mount Vernon, his 8,000-acre plantation. Absent from this precise, mathematical view is the stark reality of the 317 enslaved men, women, and children who worked within the boundaries of this map.

← *Geological and Geographical Atlas of Colorado and Portions of Adjacent Territory*, 1877.
F. V. Hayden (1829–1887)

Large-scale 19th-century surveys of the American West, commissioned after the Civil War, sought to map a railroad route across the country and identify natural resources for monetary gain. Volumes of scientific reports resulted, as did giant atlases depicting sediment types and mineral deposits in brilliant colorful swaths.

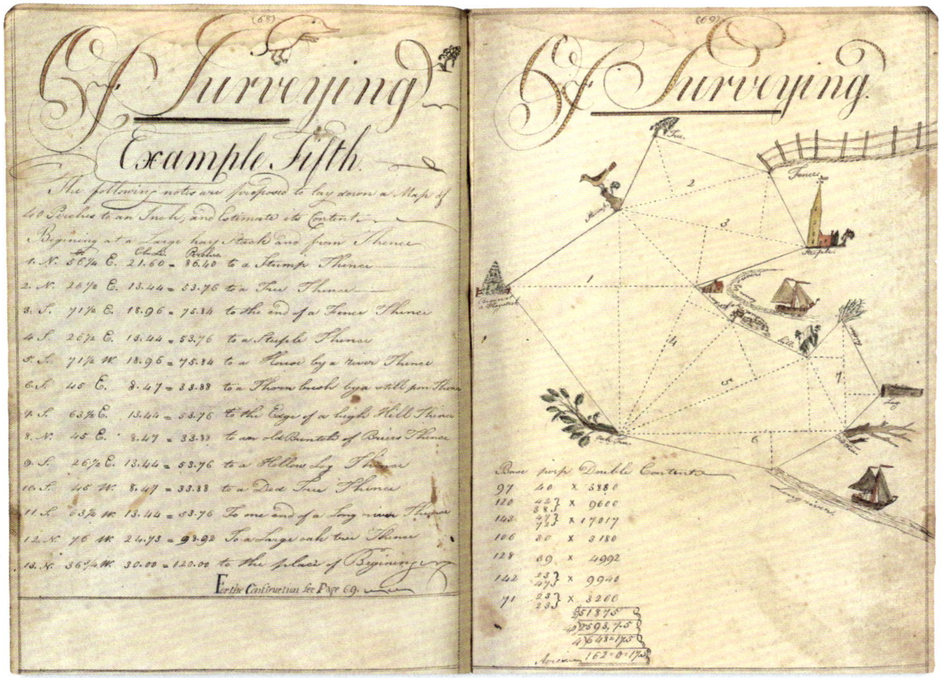

An 18th-century Delaware boy named Joseph Hunt penned a surveying exercise for school. Pupils 58 in early North America commonly made ciphering books to learn practical, everyday math. Joseph inscribed rules, problems, and corresponding solutions throughout the book in an elaborate, calligraphic hand; penmanship was another aspect of his instruction. Here he calculates acreage (called "perches") on a quirky parcel of land bounded by a haystack, a hill, a church steeple, several trees, and a hollow log.

← Arithmetic and ciphering notebook, 1798.
Joseph Hunt

Japanese surveyors in the 19th century used compasses, measuring ropes, and visual triangulation to map terrain with precision. This text emphasizes mathematical fundamentals and direct observation while cautioning against unfamiliar technologies like the octant, a maritime instrument seen as ill-suited for land.

→ *Ryōchi zusetsu* (Diagrams of surveying), 1852.
Hasegawa Hiroshi (Zenzaemon) (1782–1839)

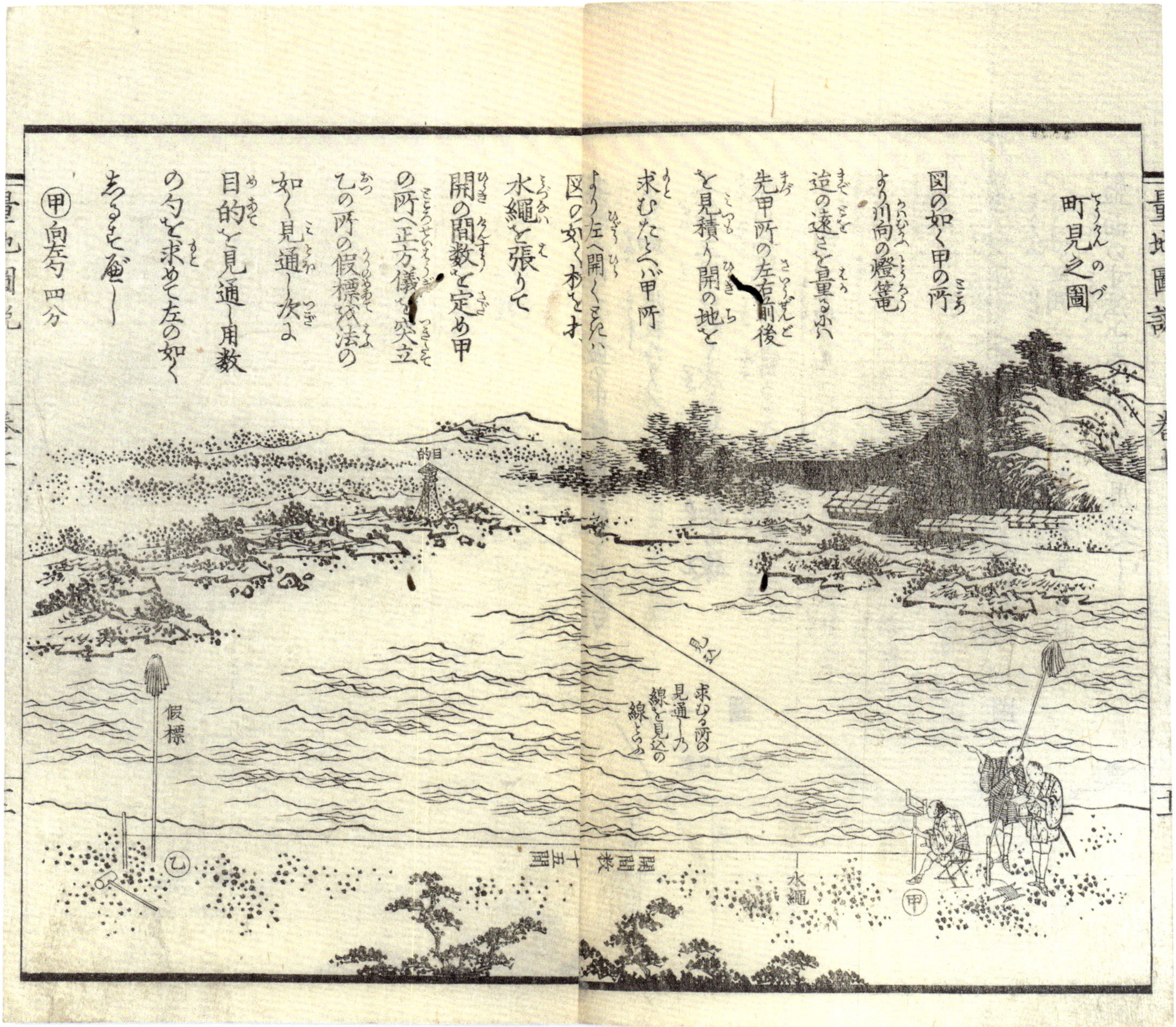
量地圖説
町見之圖
図の如く甲の所
より川向の燈籠
迄の遠さを量るハ
先甲所の左右前後
を見積り開の地を
求むたとへバ甲所
より左へ開くとなれハ
図の如く杭を打
水縄を張りて
開の間数を定め甲
の所へ正方儀を突立
乙の所の假標も法の
如く見通し次よ
目的を見通し用数
の勾を求めて左の如く
甲向左勾 四分
目的
見込
求むる所の見通しの線と見込の線
假標
十五間
開間数
水縄
甲
乙

Lay of the Land

→ *Astronaut and geologist Harrison H. Schmitt working on lunar surface during Apollo 17 mission,* 1972. Eugene A. Cernan (1934–2017)

↓ Photograph album of the U.S. Geological Survey of the Territories, ca. 1868–77. William Henry Holmes (1846–1943), compiler; William H. Jackson (1843–1942), photographer

In the 19th century, the United States funded a series of surveys to understand and settle the American West. Gargantuan undertakings involving scientists, photographers, artists, and guides, the expeditions yielded images and data from major geologic sites, including what would eventually become Yellowstone National Park in 1872. A century later, space exploration involved mapping and geology as well. Between 1962 and 1972, six Apollo lunar missions brought back 900 pounds of core samples, surface fragments, and rocks to study and test.

☽ 88

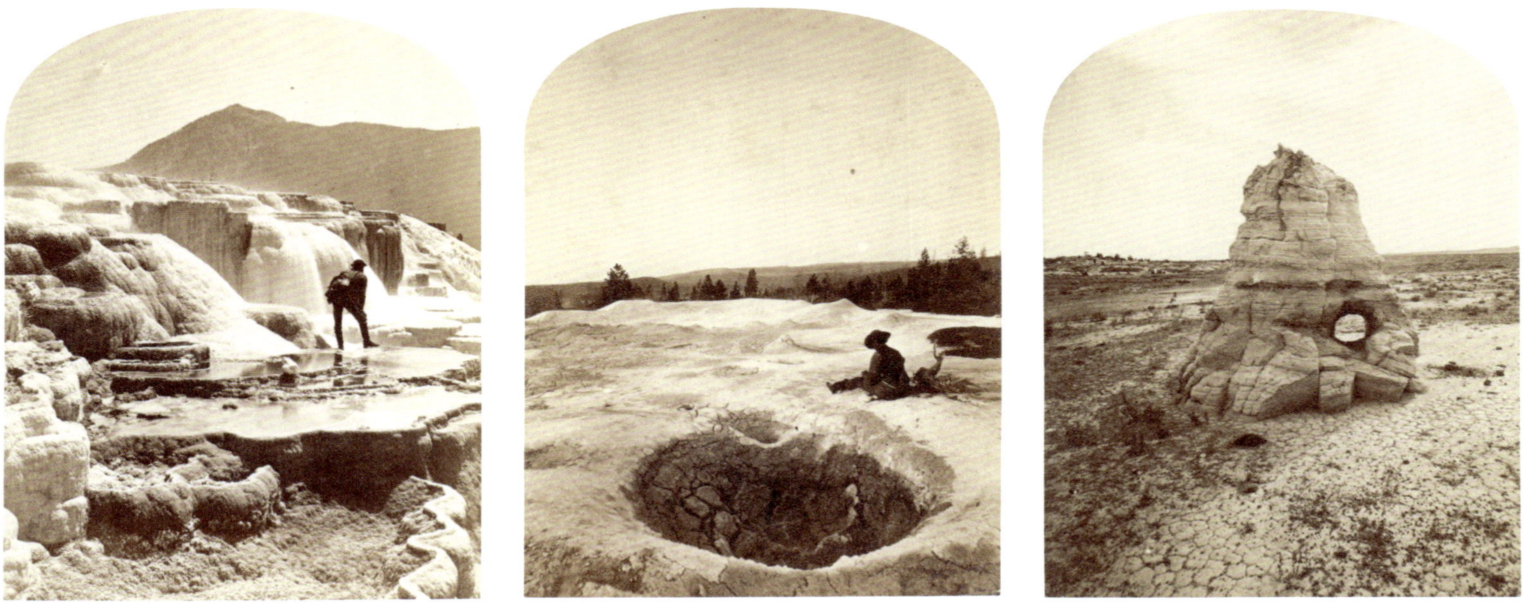

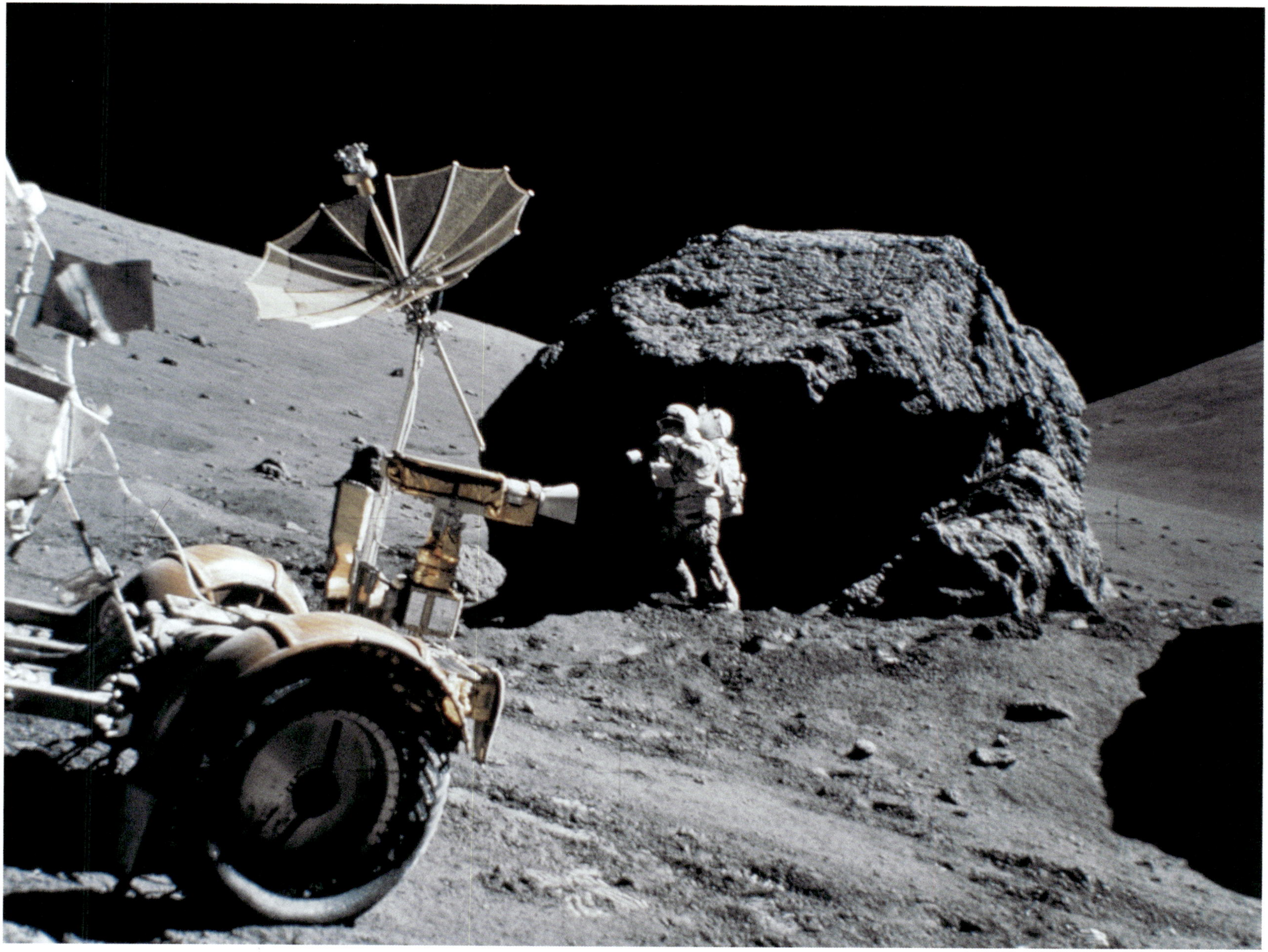

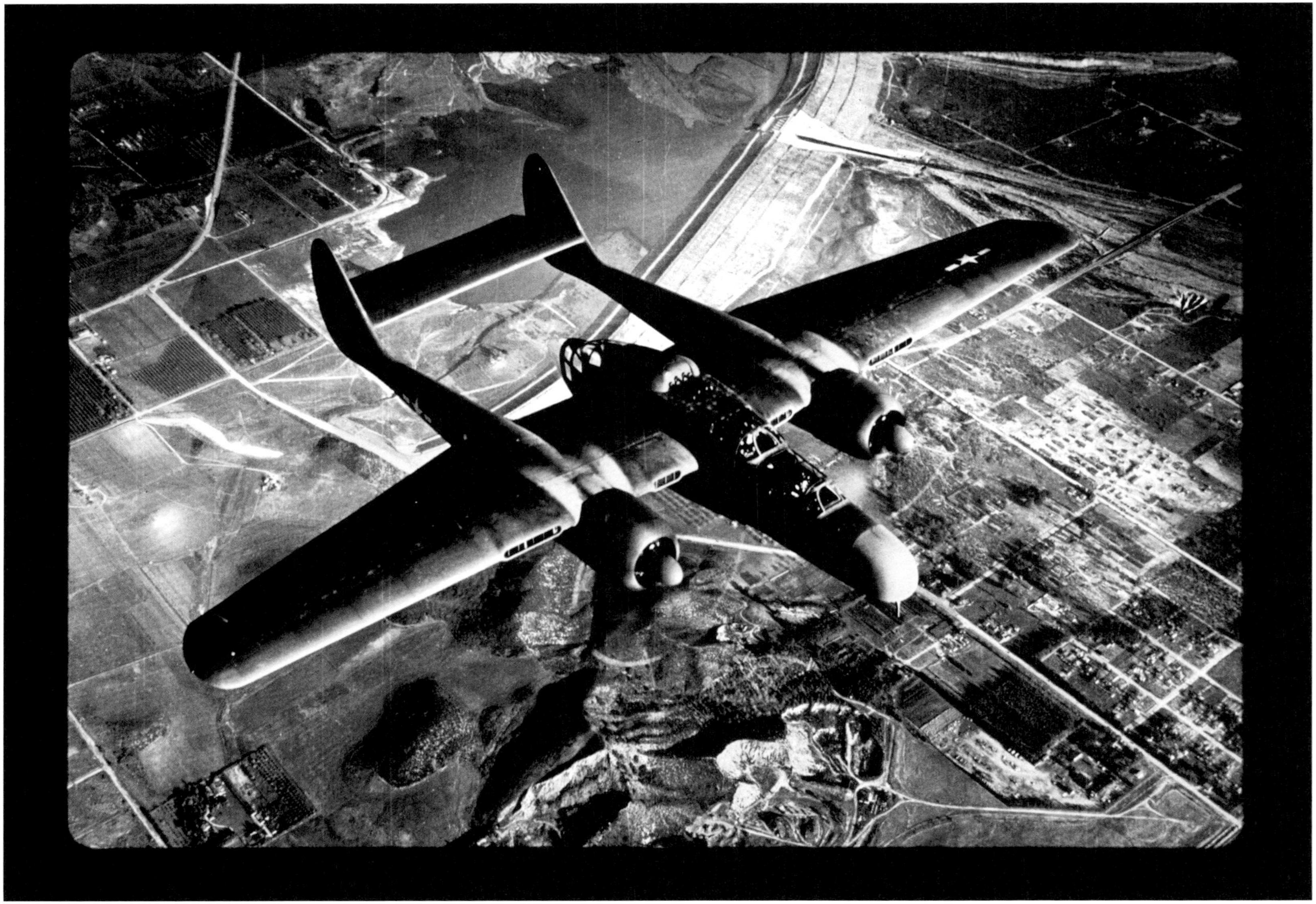

↑ *Northrop P-61 Black Widow*, ca. 1945. Northrop Grumman Corp.

Surveillance

Sophisticated aerial cameras and aircraft provided powerful surveillance capabilities during World War II. The Aerial Photographic Interpretation Detachment (APID) of the US Army examined millions of photographs tracking Germany's movements on the ground. The images, they wrote, disclosed the "enemy's cherished secrets" and provided vital assistance to the Allied campaign. Developed beginning in 1941, Northrop's P-61 was the first effective night fighter in the American military. Dubbed the "Black Widow" for its deadliness and stealth, the big plane was designed to fly for prolonged stretches without refueling and in all weather. It could loiter undetected and, according to one advertisement, "pulverize anything that flies."

54

68–71

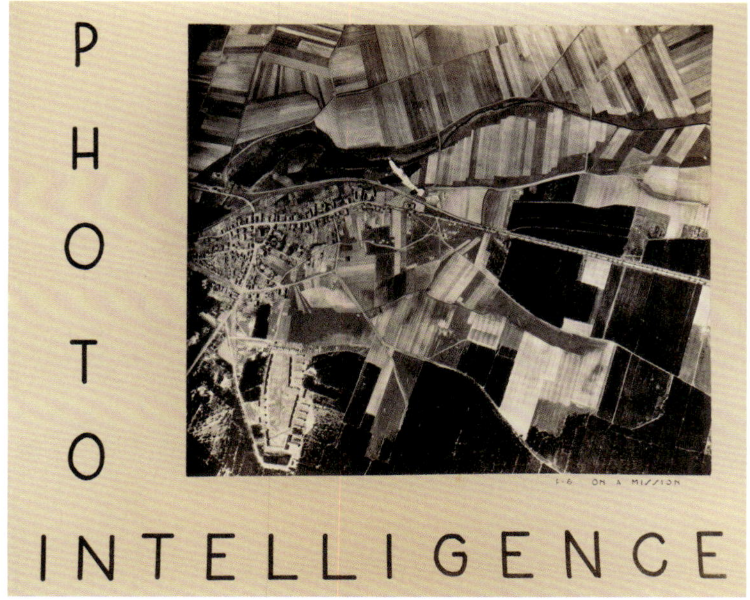

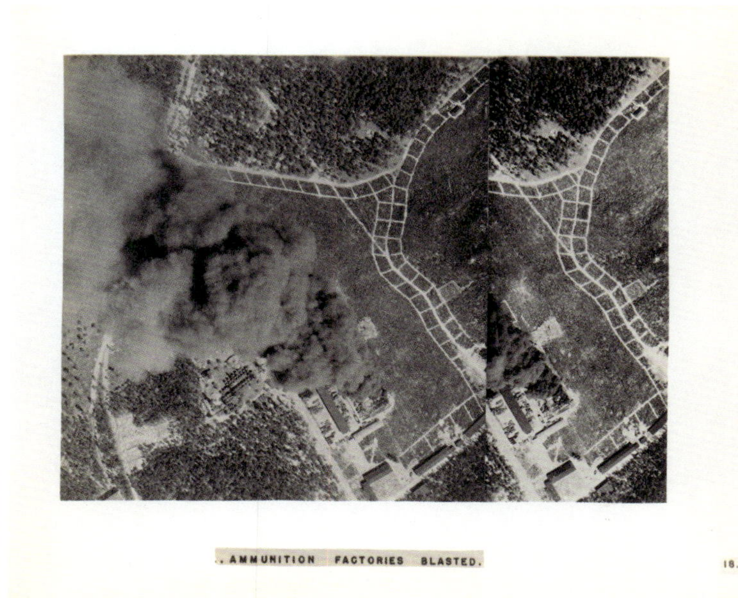

→ ↪ *Photo Intelligence*, ca. 1945. US Army Photographic Interpretation Detachment

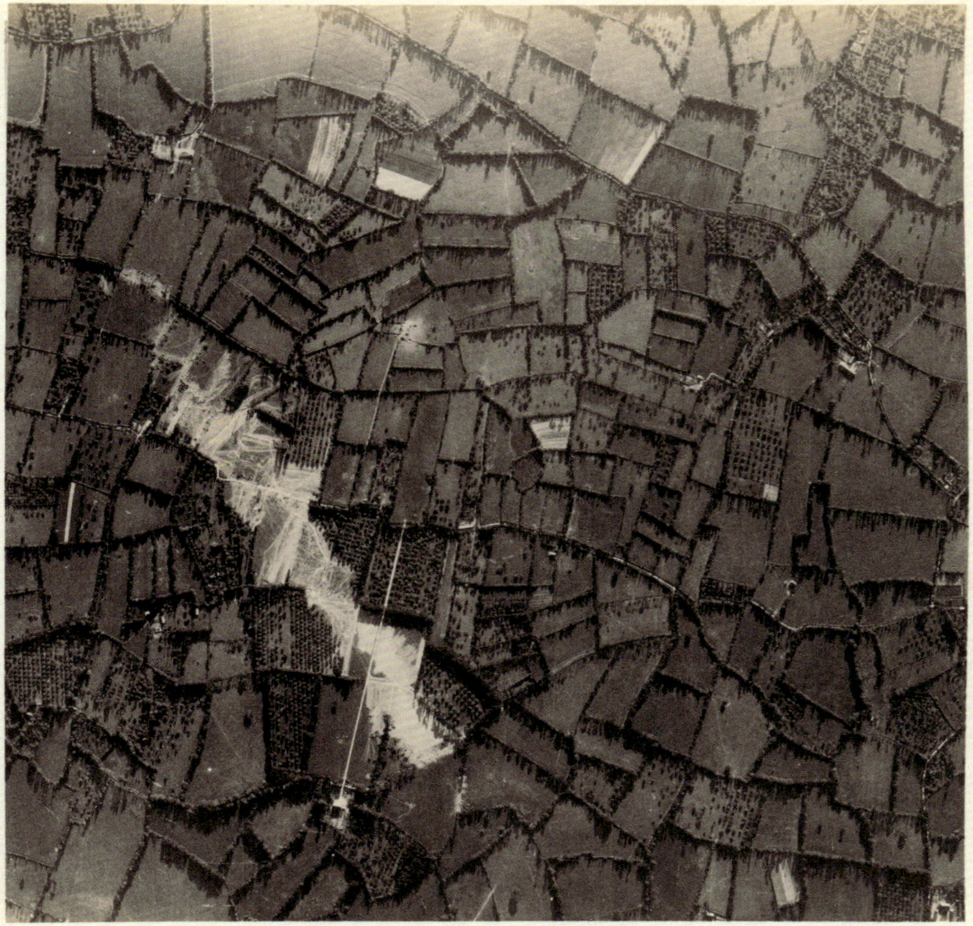

...WE CUT THROUGH THE HEDGEROWS OF NORMANDIE... 3.

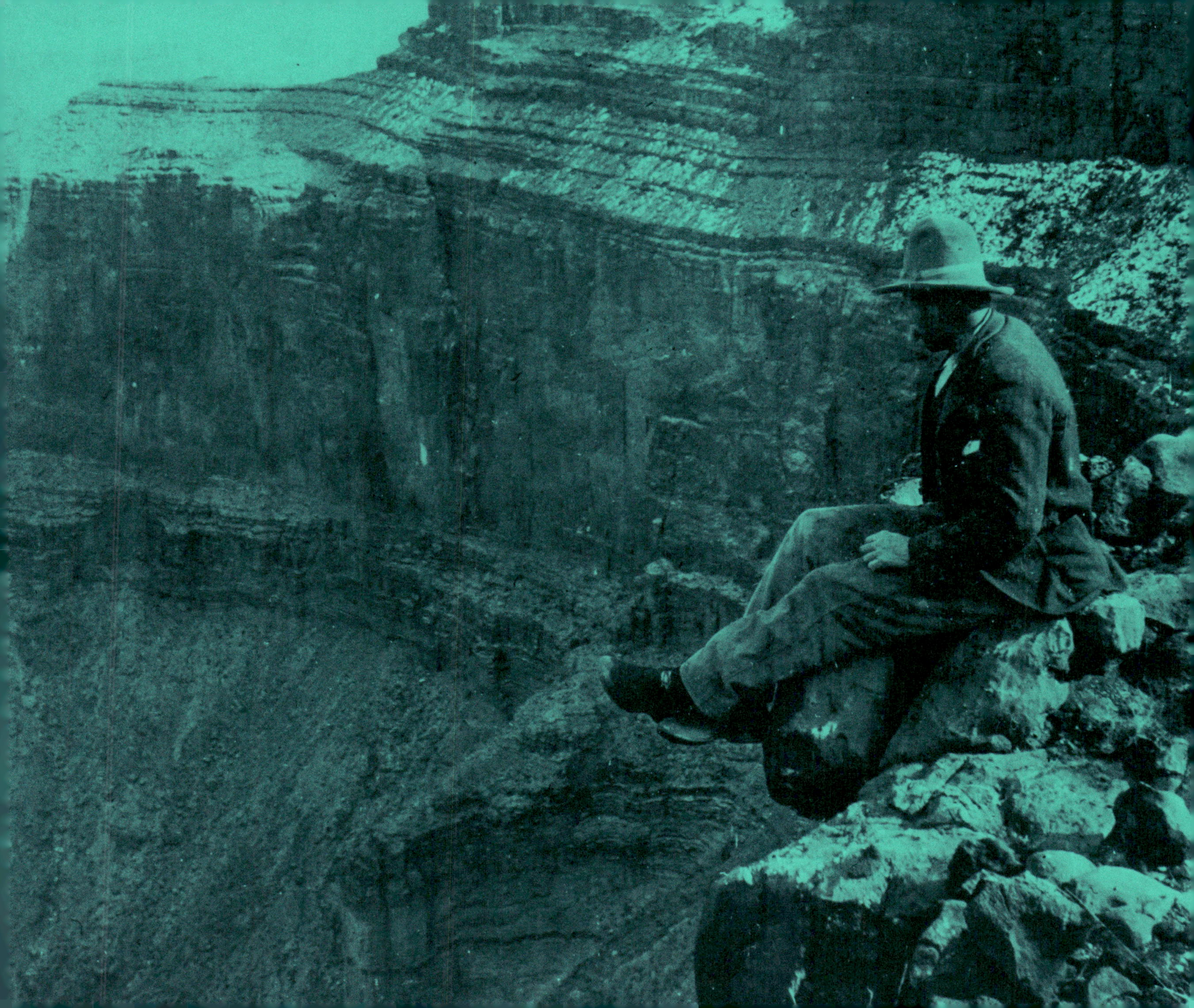

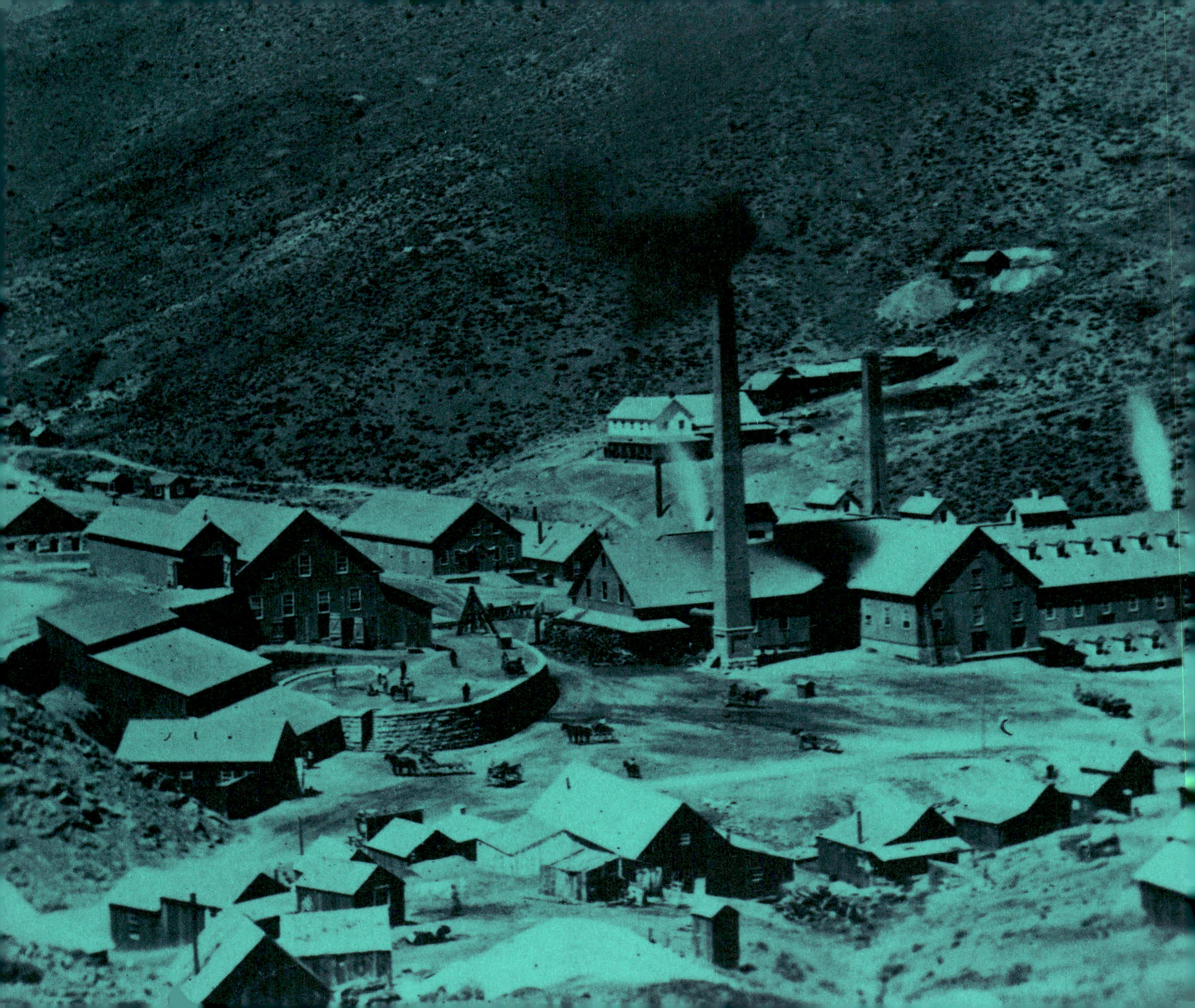

Claim

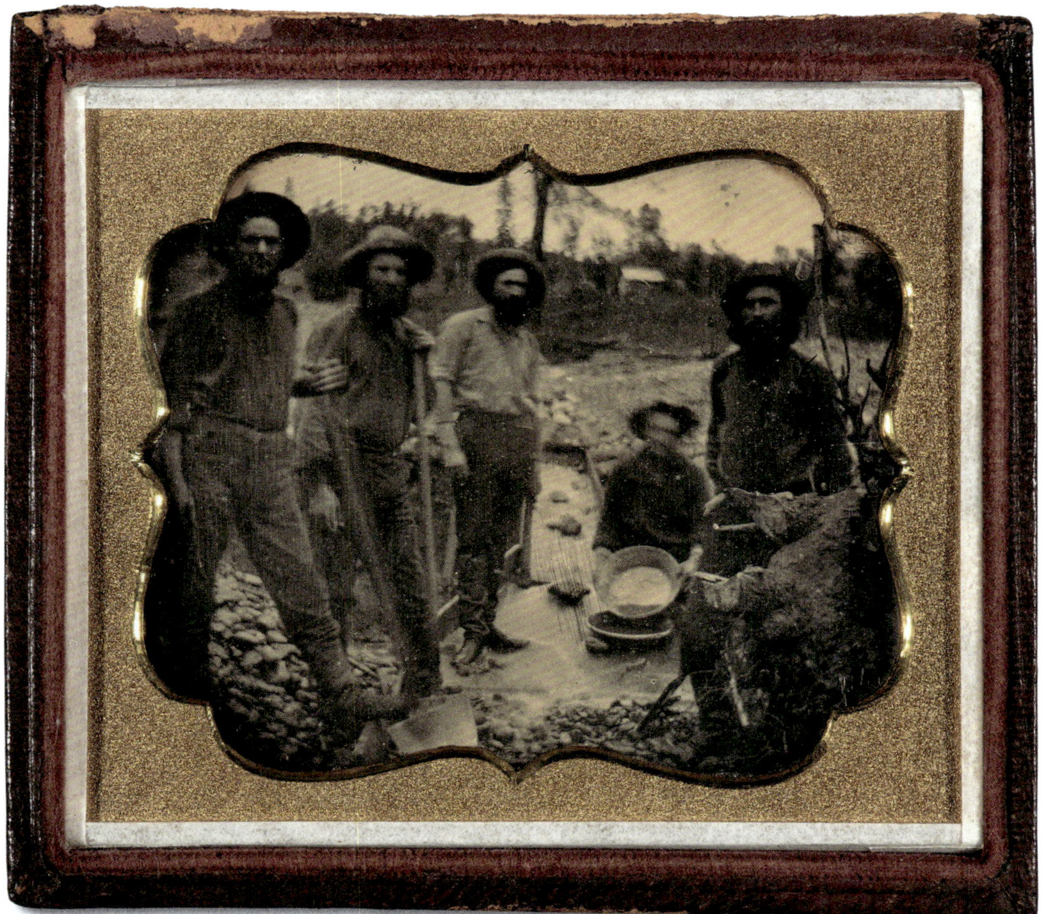

Mining in the West

89

With the 1848 discovery of gold at Sutter's Mill in California, the world rushed to the Sierra Nevada clutching shovels and pans. When the hordes exhausted the region's rivers and streams, they burrowed deep into hillsides seeking precious ore. Silver soon drew miners to other sites of grand expectation in the West, where, deep in subterranean tunnels, many met terrifying deaths by fire, asphyxiation, and cave-ins. Early mining in the American Southwest created a patchwork of claims as well as overnight cities powerless against cycles of boom and bust. What remains long after all the digging, drilling, blasting, and scraping are poisoned watersheds and scarred terrain.

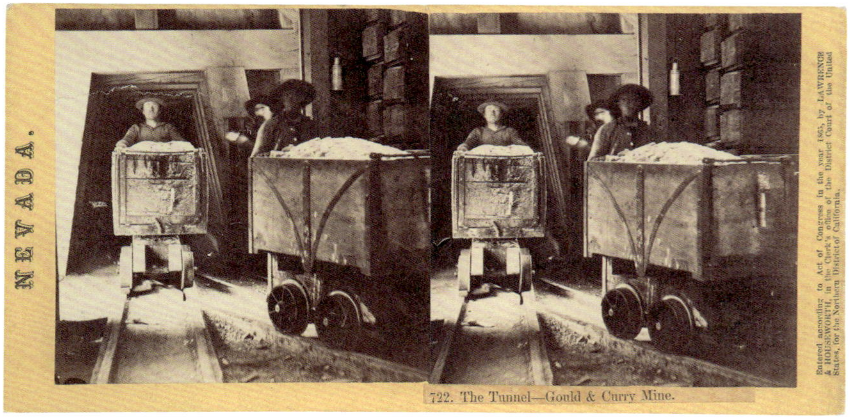

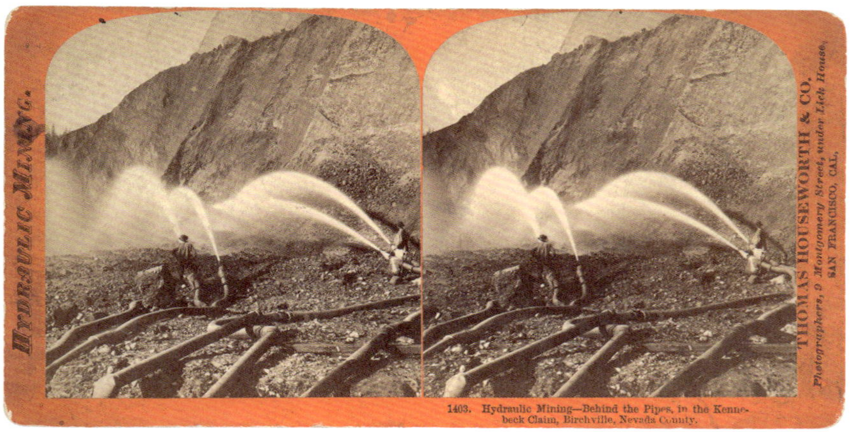

↩ Daguerreotype of five miners in California with shovels and pans, ca. 1848–50s

← (TOP) Stereograph of *The Tunnel, Gould & Curry Mine, Nevada*, ca. 1865. Lawrence & Houseworth (1851–68); (BOTTOM) Stereograph of *Behind the Pipes in the Kennebec Claim, Birchville, Nevada*, ca. 1870s. Thomas Houseworth & Co. (1868–90)

→ *Mining on the Comstock* broadside, 1876. J. B. Marshall, publisher; T. L. Dawes, artist

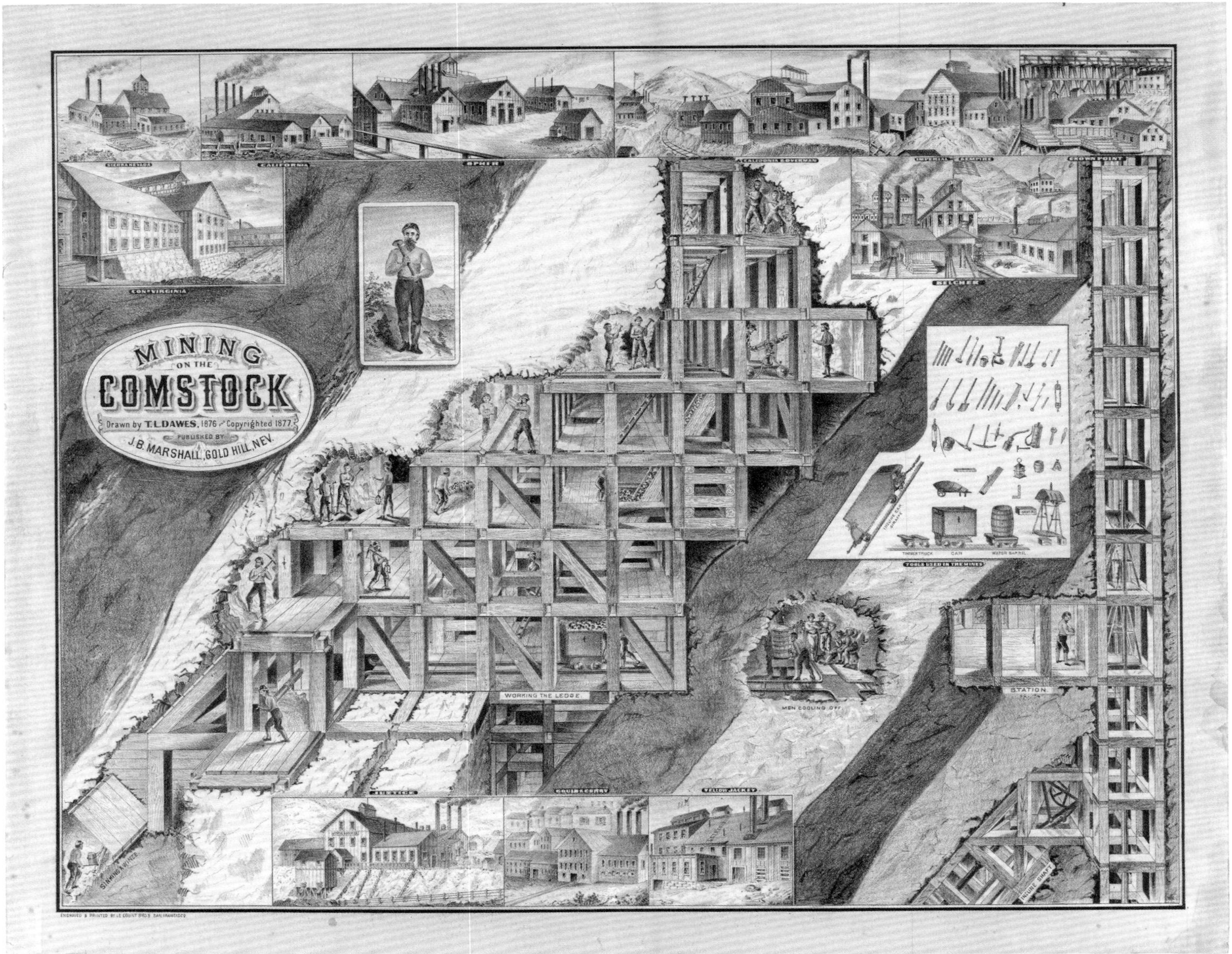

THE SPEARHEAD GOLD MINING CO. REORG. HAS TAKEN OVER A PROPERTY IN THE COMSTOCK DIST.

→ Details from *'Mongst Mines and Miners, or Underground Scenes by Flashlight*, 1893. J. C. Burrow (1852–1914), author and photographer; William Thomas (1857–1931), author

↓ *Diannan kuangchang tulüe* (Illustrated account of mining in Diannan), 1844. Wu Qijin (1789–1847)

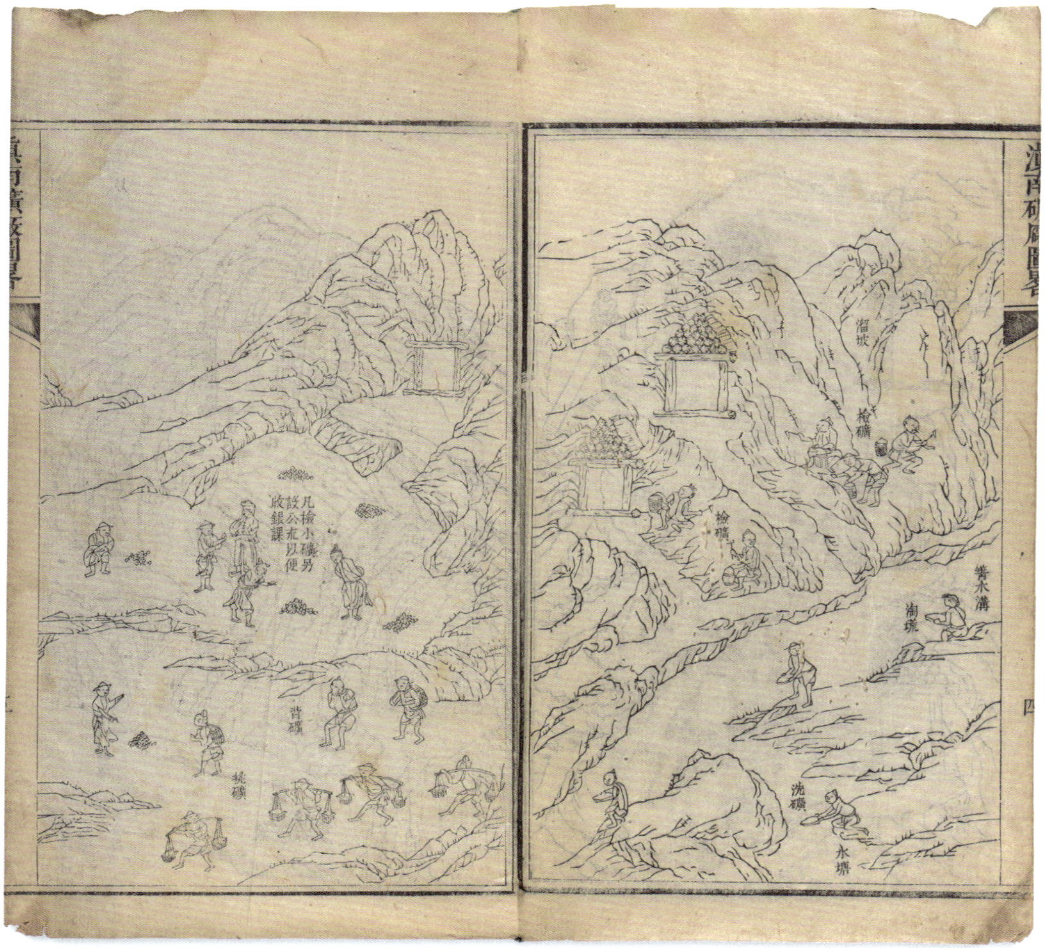

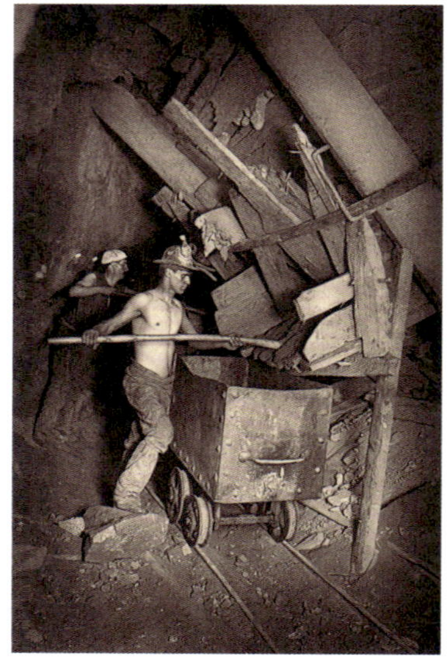

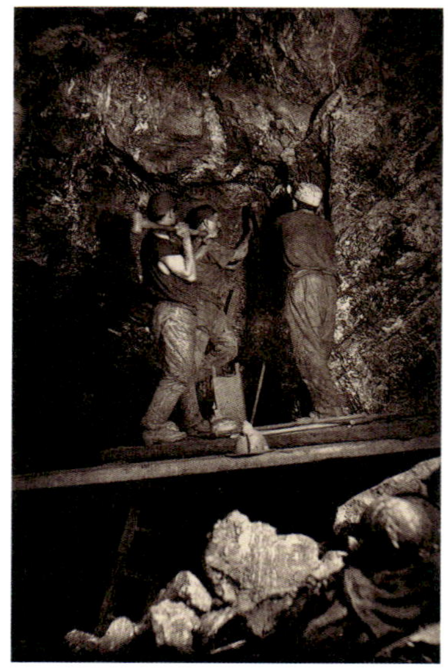

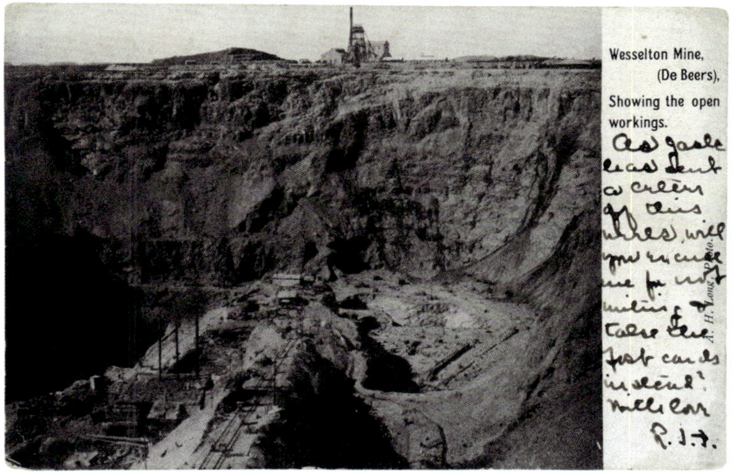

↑ Postcard of Wesselton diamond mine in Kimberley, South Africa, operated by De Beers Consolidated Mines Ltd., ca. 1906

↪ *Kinginzan funai shikioka kasegikatazu* (Illustrations of operations inside and outside of gold and silver mines), Sado Island, Japan, after 1788

INSIDE COVER

A visual drama of 18th-century Japanese mining unfurls over an astonishing 64 feet. Two hand-painted scrolls lead the viewer on a tour of mineral-rich Sado Island, beginning in dank underground tunnels and ending at a bustling mint. Along the way, the traveler encounters the tools, techniques, processes, and laborers—of every age, class, and rank—transforming silver and gold into precious coins. Sado's mines once supplied 10 percent of the world's gold. They closed in 1989.

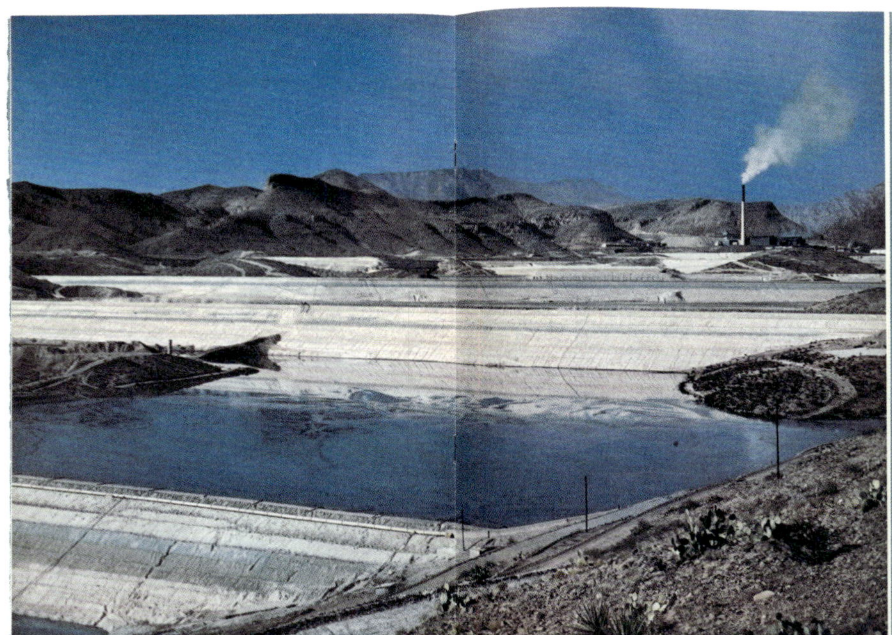

→ Annual Report, 1950. Phelps Dodge Corp.

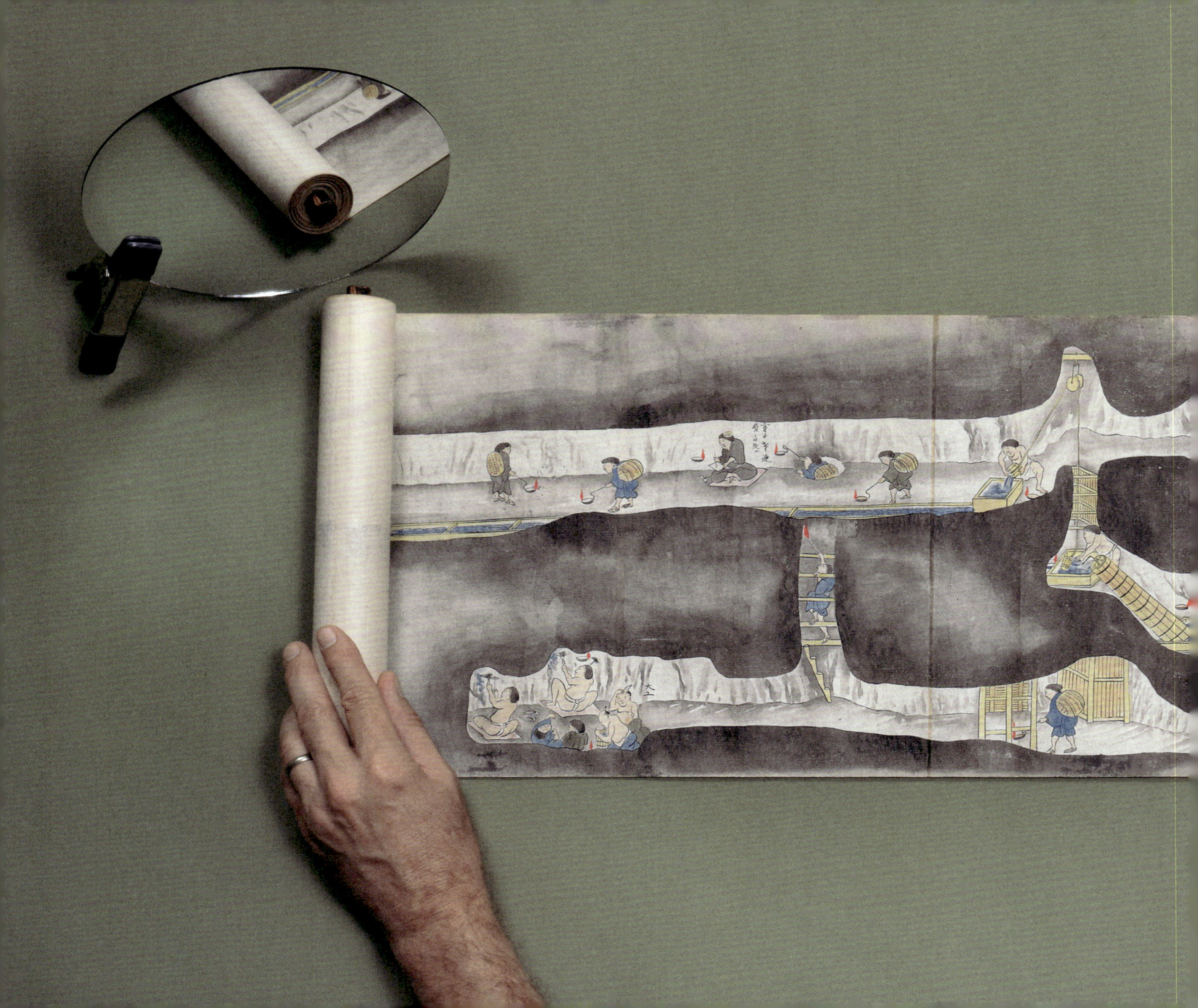

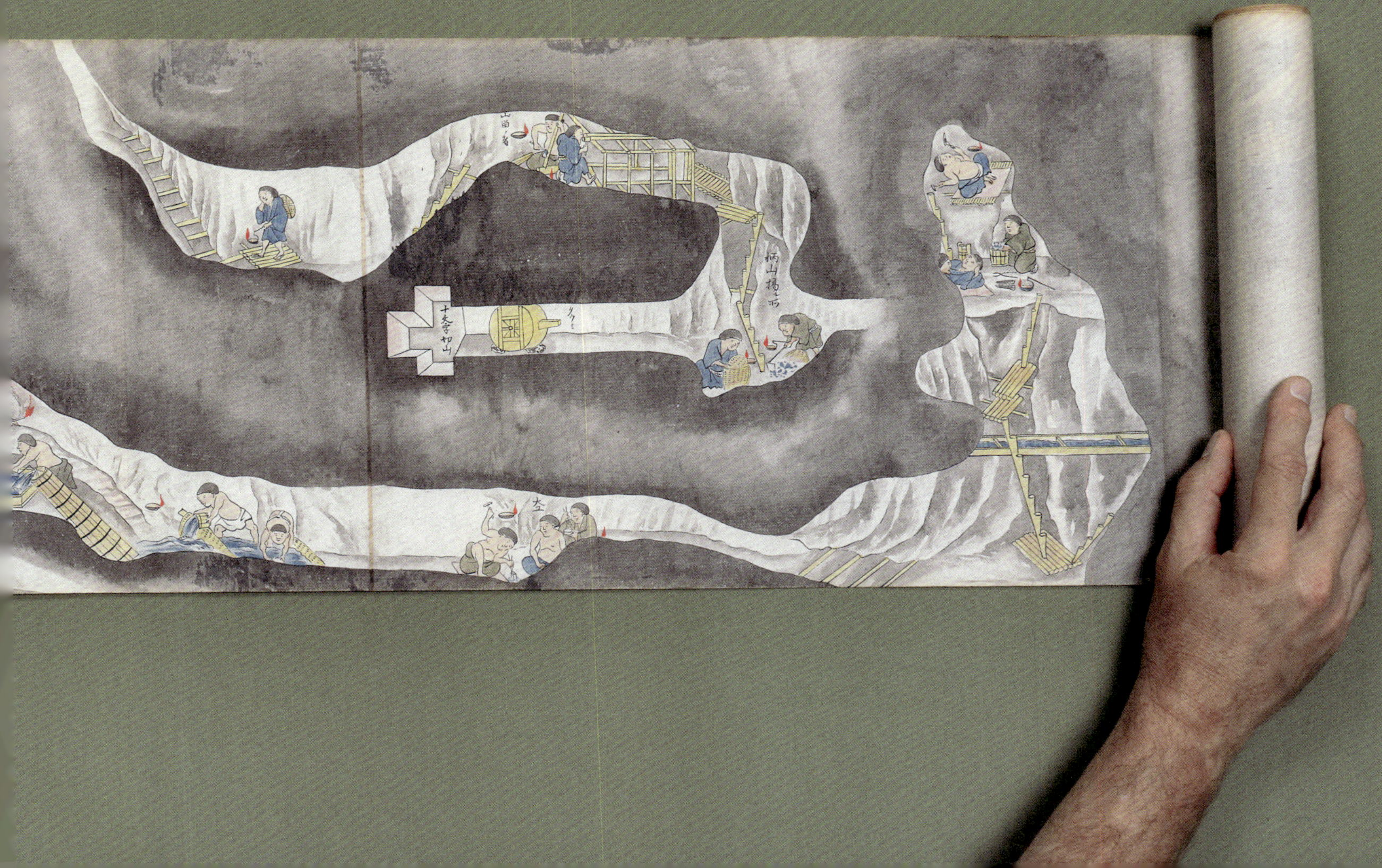
十文字切山

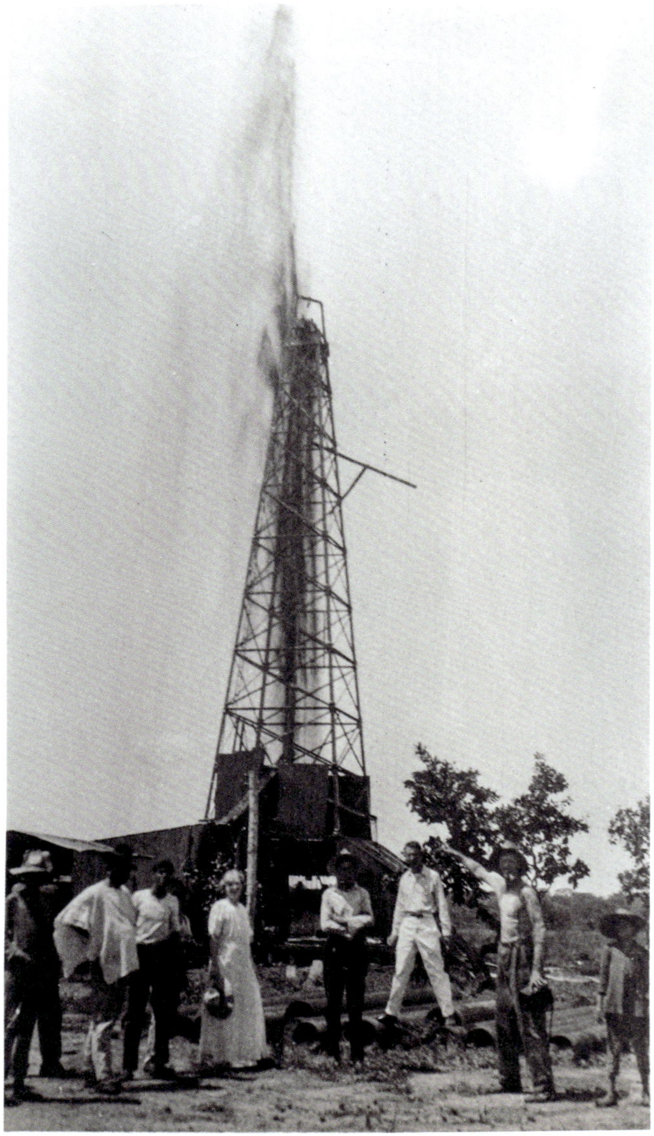

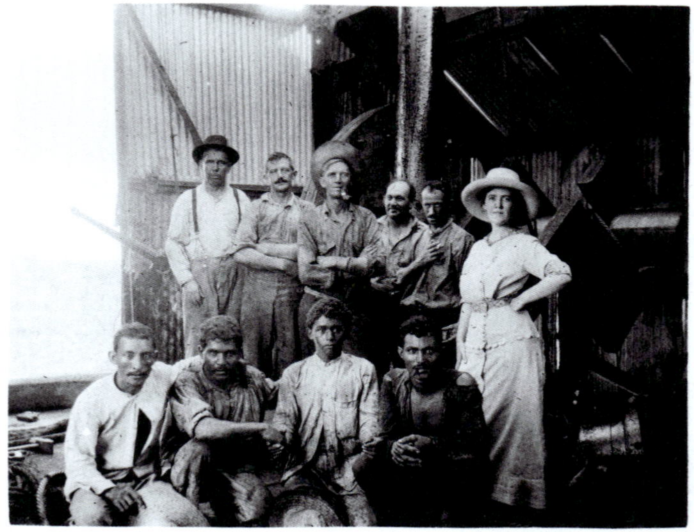

No event in Venezuela can be separated from oil....All aspects of the Venezuelan economy are the legitimate or bastard children of that substance that irrevocably stained our history.

—Domingo Alberto Rangel, *La moneda ladrona* (The thieving currency), 1964

Oil in Venezuela

Between 1911 and 1916, Ralph Arnold, a California-trained geologist, led a Venezuelan oil hunt at a US company's request. The mission? To find "black gold" and pump it out of the ground. Arnold was ordered to claim every acre of petroleum-soaked land. He and his team of more than 50 scientists traversed hundreds of difficult miles across swamps, deserts, and tropical forests, interviewing rural laborers and gauging the terrain. He kept precise photographic and written records to generate reports. One 1912 notebook contains sketches depicting potential reserves at "El Mene" (a gooey, black seepage) in western Venezuela on Lake Maracaibo's shore. The site, Arnold concluded, was destined to yield gushers of oil. The president of Royal Dutch Shell agreed, paying millions in a speculative maneuver that made the company and its shareholders very rich.

⩰ 78

↖ *Interior view of Zumba well, showing drilling crew and Mrs. Stokes, Mene Grande, Zulia,* 1914. John A. Stokes

← *Looking southwest at Zumaya No. 1 during its period of maximum flow, Mene Grande, Zulia,* 1914. John A. Stokes

→ *Resumen de actividades* (Annual report), 1953. Compañia Shell de Venezuela (Venezuela Shell Company)

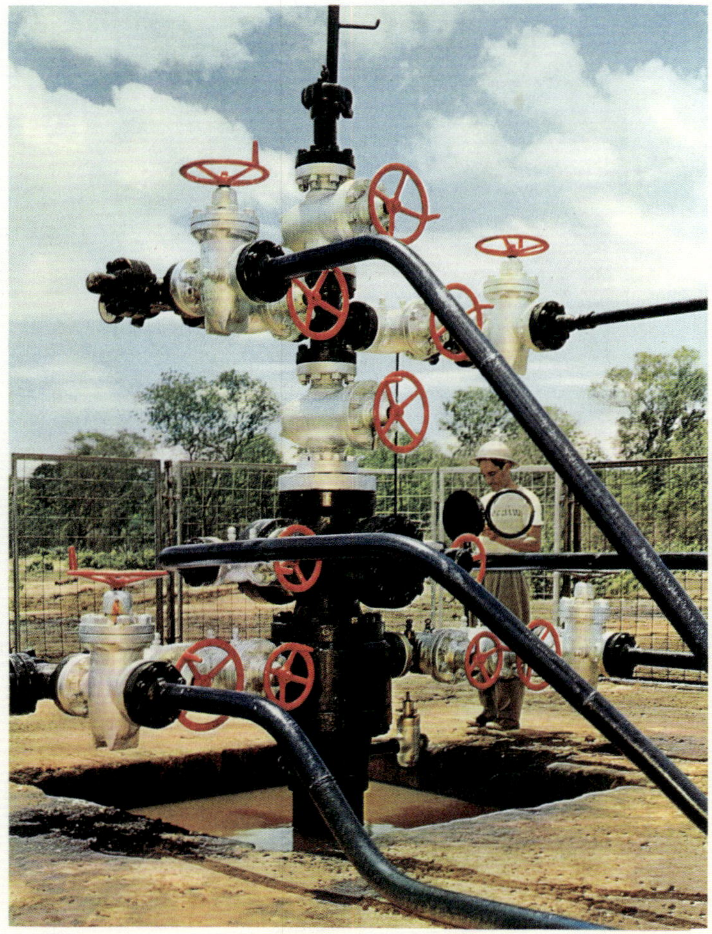

Una cruceta o "árbol de navidad"

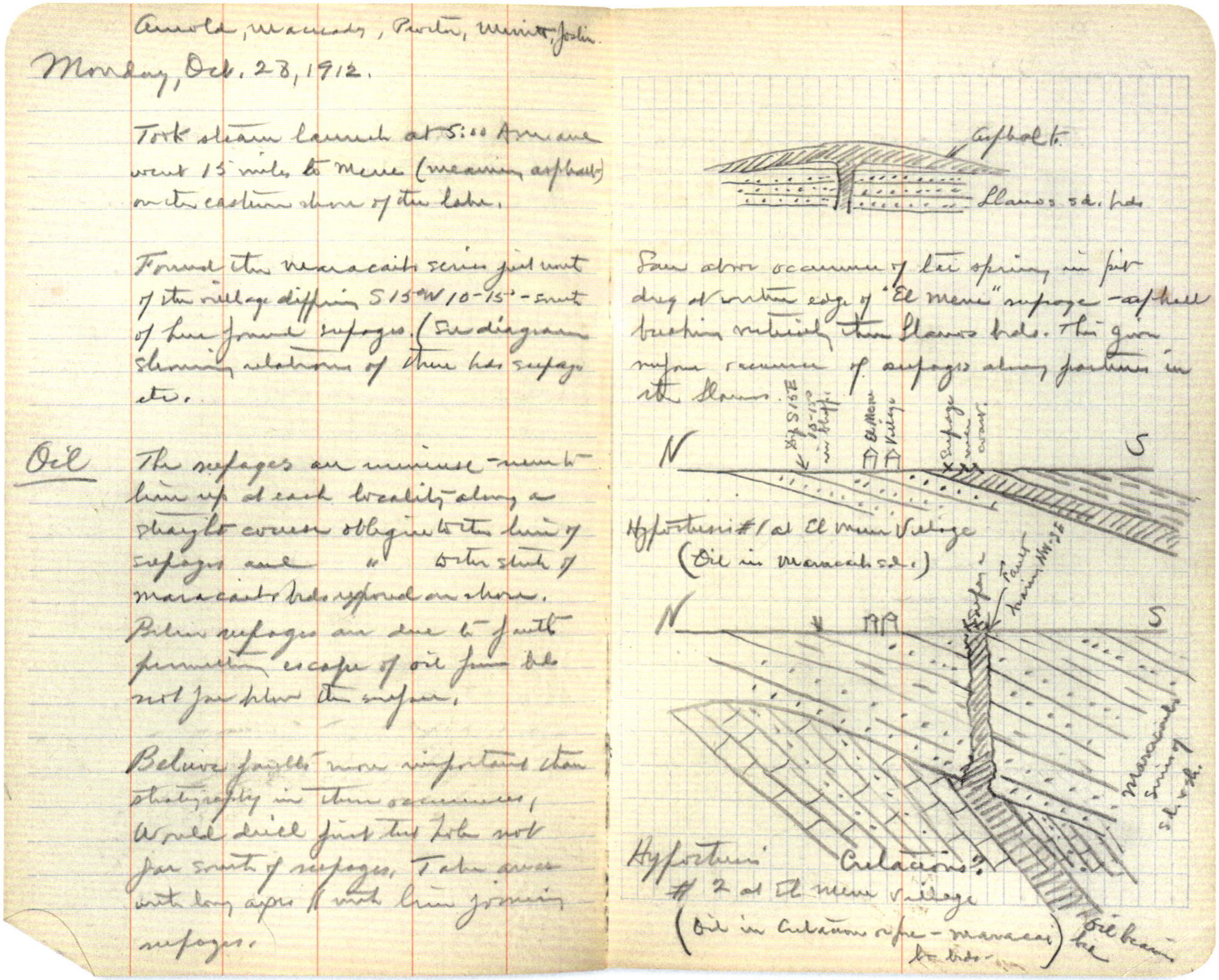
Arnold, Macready, Porter, Merritt, John.

Monday, Oct. 28, 1912.

Took steam launch at 5:00 Am and went 15 miles to Mene (meaning asphalt) on the eastern shore of the lake.

Found the Maracaibo series just west of the village dipping S15°W 10-15° - south of line found seepages. (See diagram showing relations of these beds seepages etc.

Oil

The seepages are numerous - seem to line up at each locality along a straight course oblique to the line of seepages and " strike of maracaibo beds exposed on shore. Believe seepages are due to faults permitting escape of oil from beds not far below the surface.

Believe faults more important than stratigraphy in these occurrences, Would drill first test hole not far south of seepages.

Saw above occurrence of tar spring in pit dug at western edge of "El Mene" seepage - asphalt breaking vertically thru Llanos beds. This gives surface occurrence of seepages along fractures in the Llanos.

↑ Field book related to oil fields at Lake Maracaibo, Venezuela, 1912. Ralph Arnold (1875–1961)

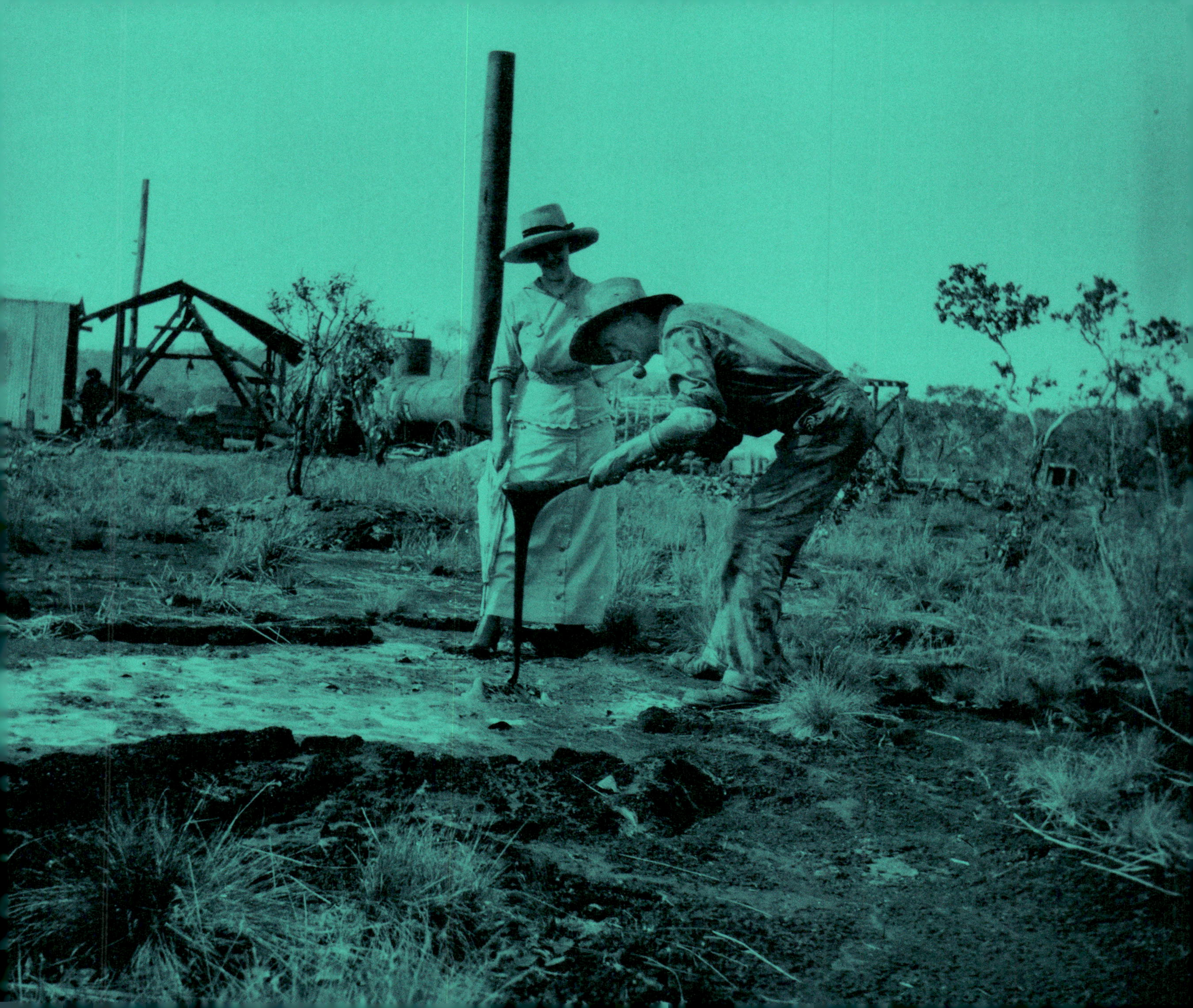

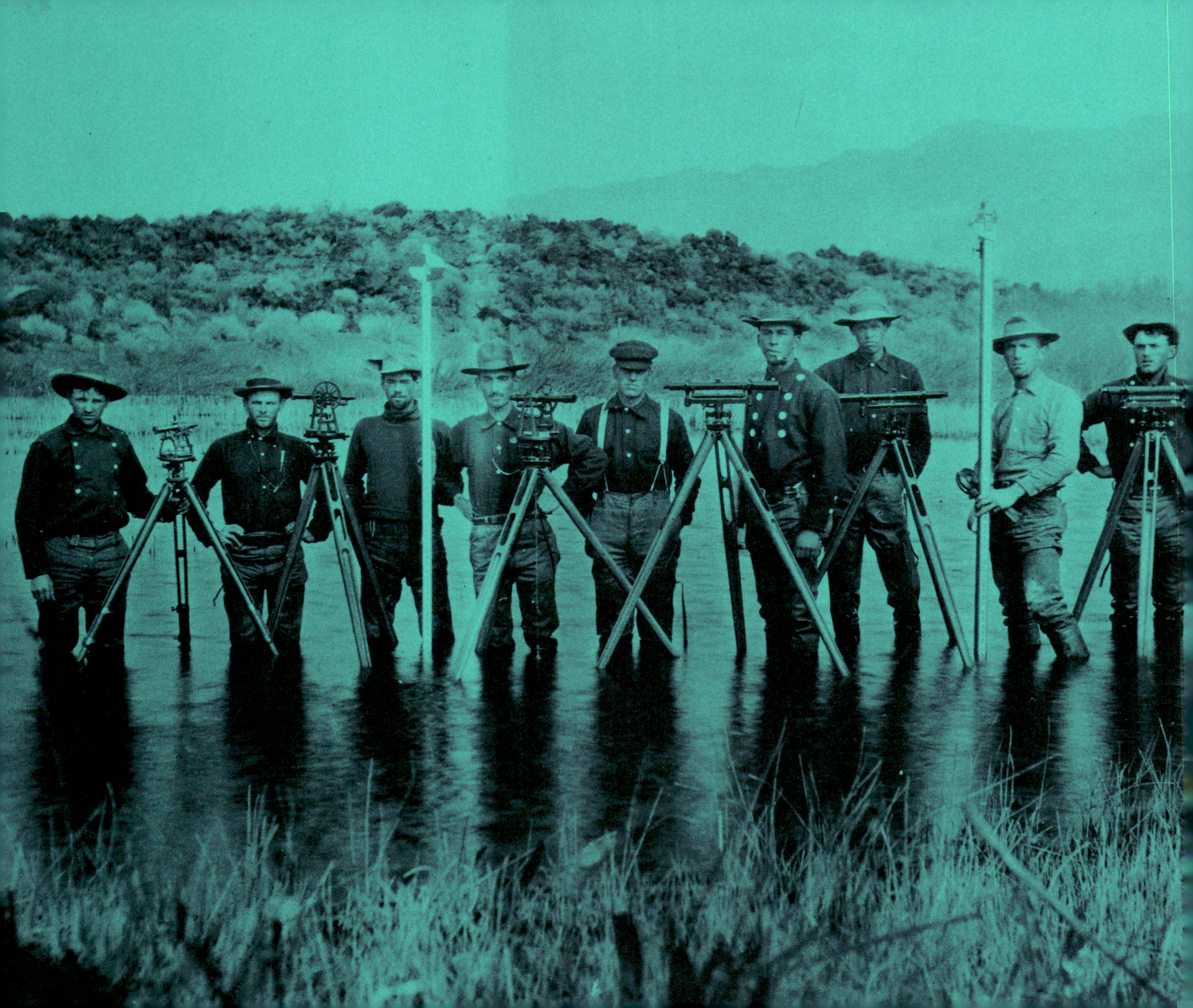

Divert

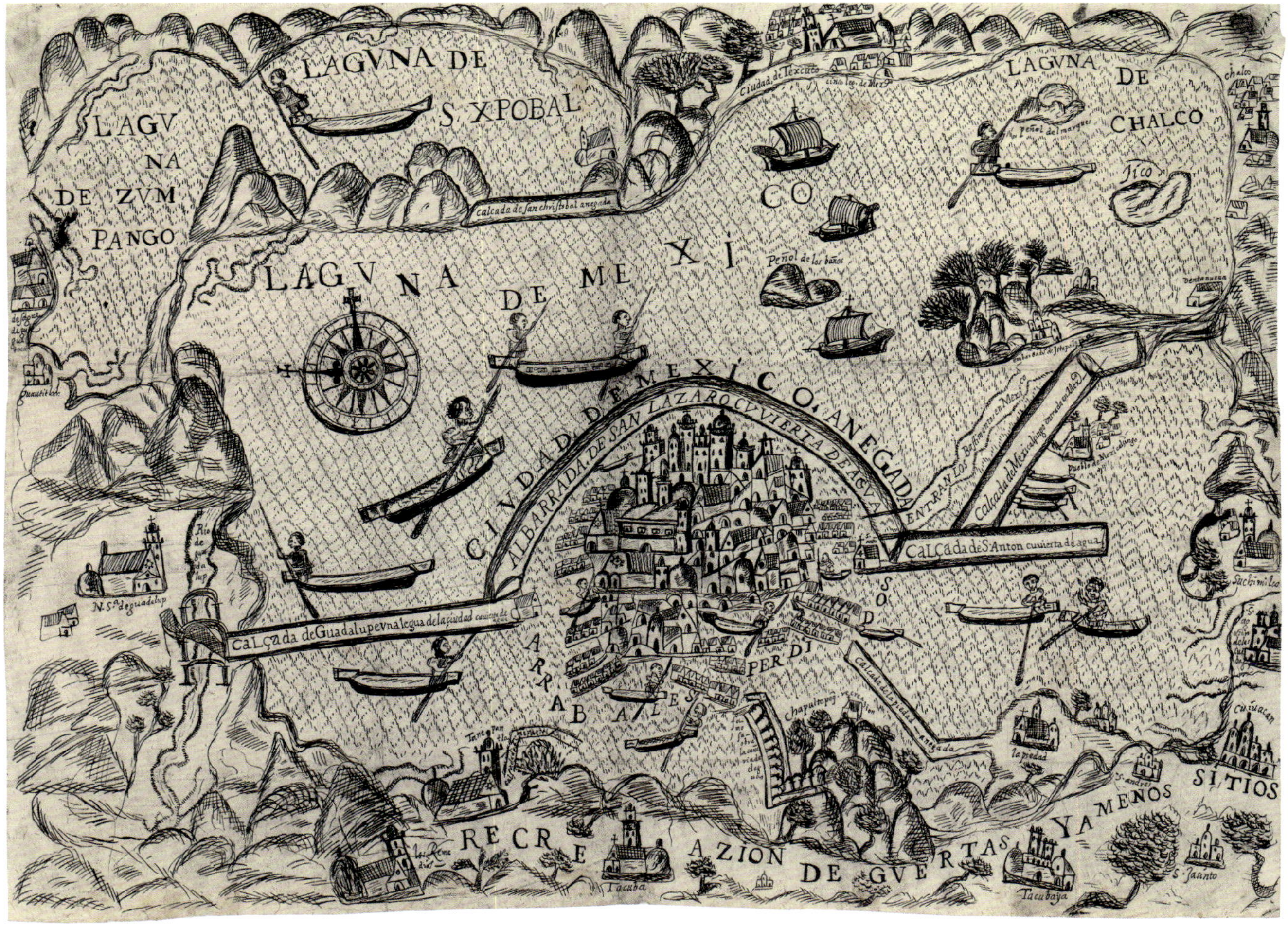

LAGVNA DE S XPOBAL
LAGV NA DE ZVM PANGO
LAGVNA DE CHALCO
LAGVNA DE ME XI CO
calcada de san christobal anegada
Ciudad de Texcuco
Peñol de los baños
Tacuba
Tacubaya
CIVDAD DE MEXICO ANEGADA
ALBARRADA DE SAN LAZARO CVVIERTA DE AGVA
CALCADA DE S. ANTON cuuierta de agua
CALÇADA DE GUADALUPE VNA LEGUA DE LA CIUDAD
ARRABALES
RECREAZION DE GVERTAS Y AMENOS SITIOS

← Hand-drawn map of the Valley of Mexico, after 1652

The Valley of Mexico once nourished a basin of interconnected lakes. Early inhabitants adapted accordingly, engineering an elaborate system of dams, levees, aqueducts, canals, even floating gardens. The ancient island metropolis of Tenochtitlan (present-day Mexico City) flourished here. Beginning in 1519, Spanish conquerors destroyed much of this hydraulic infrastructure, leaving ☽ 12
communities vulnerable to heavy rains. This map records the devastating flood of 1629, which lasted five years and claimed thousands of lives. The arch at the map's center, Albarrada de San Lazaro, marks a containment wall, and several causeways, *calçadas*, lead into the city. Nearly all are labeled *cubierta de agua* (covered in water) or *anegada* (flooded). Throughout the subsequent 400 years, the region's lakes were systematically drained.

Engineering Empires

The early 20th century saw massive engineering projects emerge around the world. These were typically imperial undertakings in colonial outposts. The construction of the Panama Canal, begun in 1881, displaced acres of water and rock to create one of the world's most impressive shortcuts. The canal promised to turn a ship's 13,000-mile, months-long transit around South America's Cape Horn into mere weeks. Initial French efforts failed. US investment followed, with the canal completed in 1914. Half a world away in Egypt, British engineers designed and built the Aswan Dam to span the Nile River. Intermittent construction over decades increased control over annual flooding, allowed for year-round water storage and irrigation, and provided hydroelectric power. An alarming decrease in river biodiversity as well as erosion and the buildup of silt are some of the dam's ecological harms.

↗ (TOP) Postcard of sailors viewing Gatun Locks, Panama Canal, ca. 1914–23. Vibert & Dixon. (BOTTOM) Postcard of men mixing concrete at Miraflores, Panama Canal, ca. 1910s. L. L. Maduro Jr.

→ *Aswan Dam heightening, Lock no. 1, Gate no. 20, 1910*

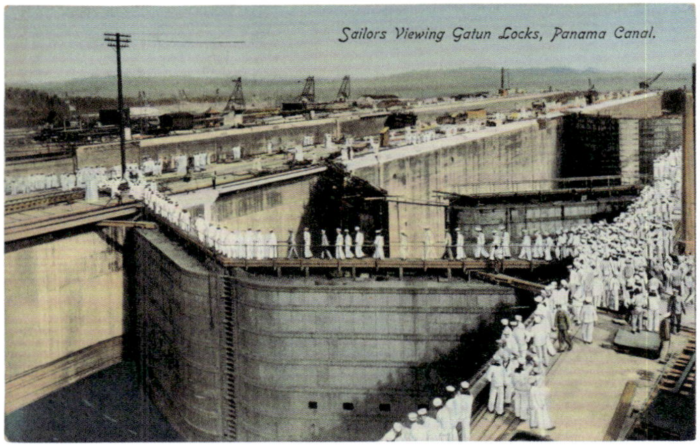
Sailors Viewing Gatun Locks, Panama Canal.

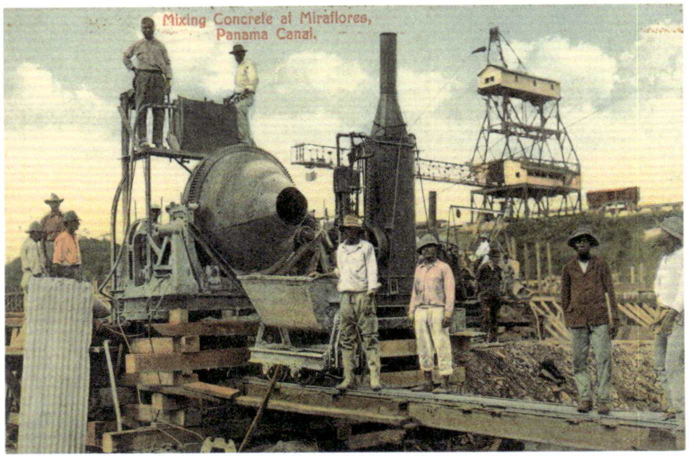
Mixing Concrete at Miraflores,
Panama Canal.

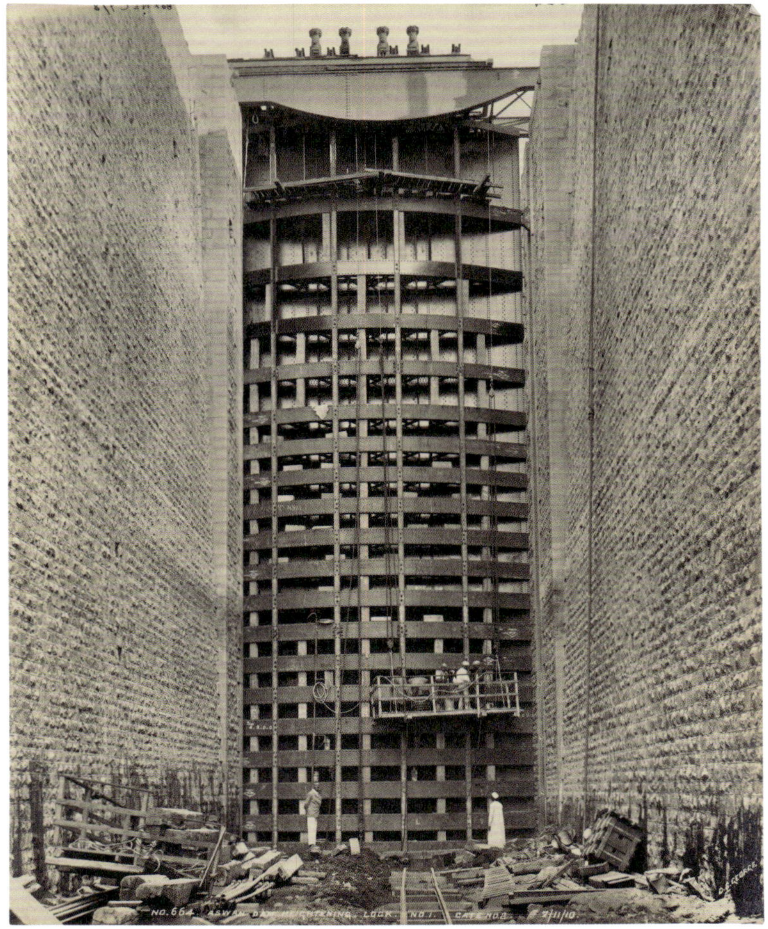
NO. 664. ASWAN DAM HEIGHTENING. LOCK. NO 1. GATE NO 2. 2/11/10.

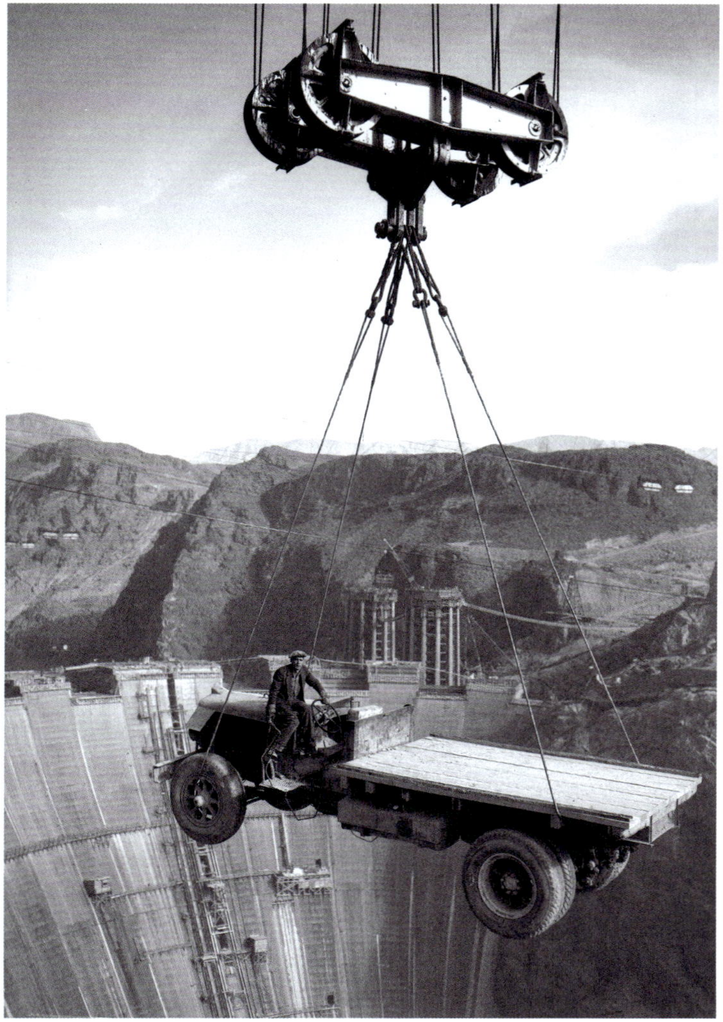

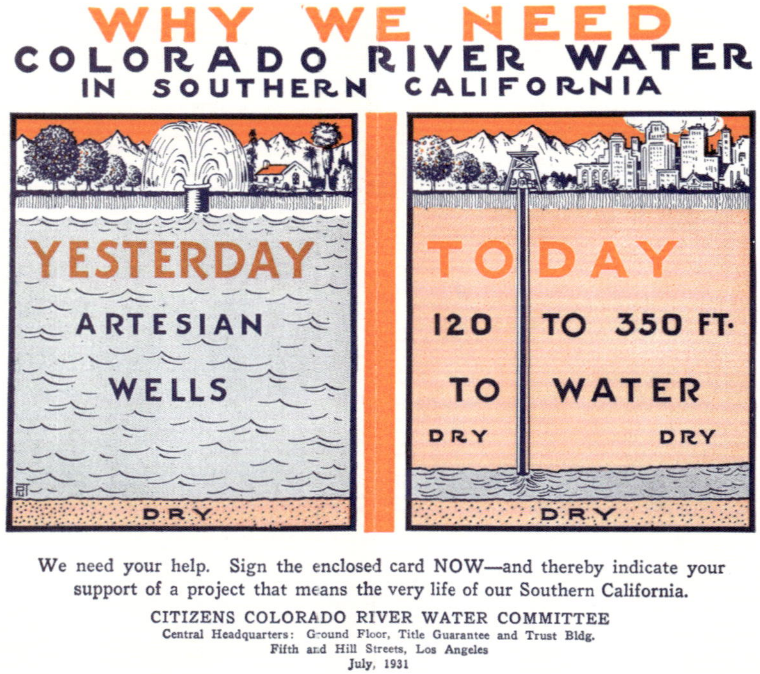

↑ Colorado River brochure, 1931. Citizens Colorado River Water Committee

← *Construction at Boulder Dam* (Hoover Dam), 1934. "Dick" Whittington Studio (1924–87)

A Great Thirst

As world populations expanded, people engineered increasingly sophisticated methods to slake a growing thirst. In the arid southwestern United States, this was particularly true; in 1913, Los Angeles drank up Sierra Nevada snowmelt via 233 miles of concrete aqueducts and gravity-fed steel tubes. The Hoover Dam, built during the Great Depression with 20,000 workers and six million tons of concrete, harnessed the Colorado River to irrigate, generate power, and supply potable water across seven western states. As global temperatures continue to rise, by the time the river reaches Mexico, it now runs mostly dry.

FRONT COVER

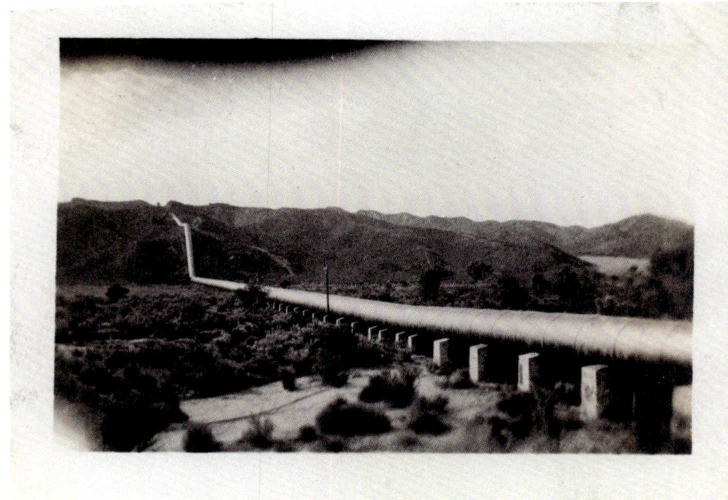

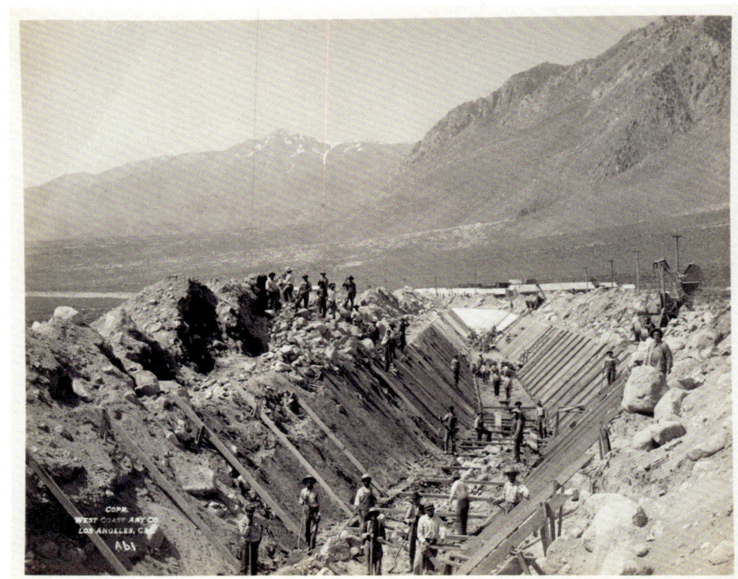

↗ (TOP) *Los Angeles Aqueduct pipeline*, ca. 1915–25; (BOTTOM) *Concrete lining Alabama Hills, Owens Valley, California*, ca. 1910. West Coast Art Co.

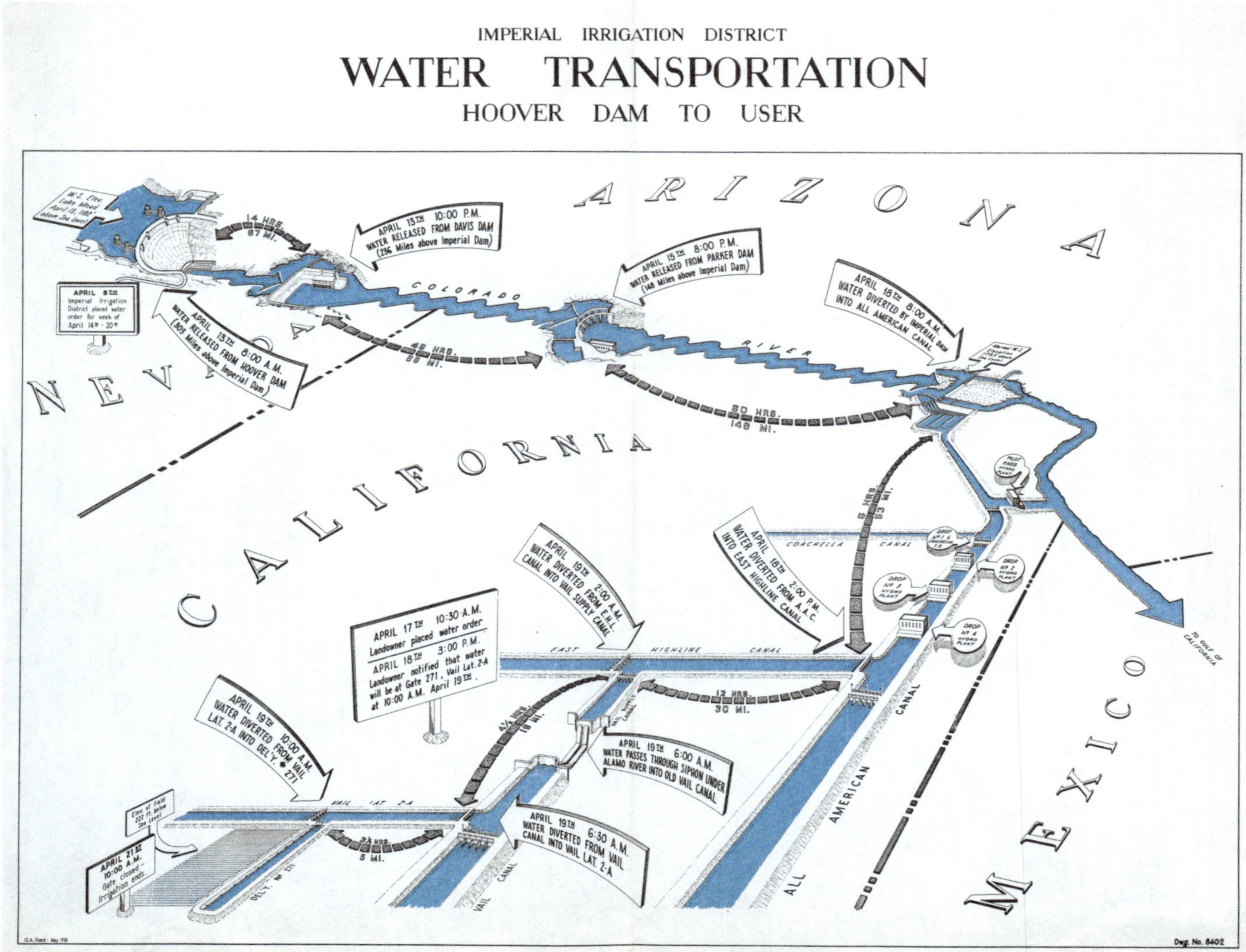

↑ *Water Transportation: Hoover Dam to User* leaflet, 1958. Imperial Water District, publisher; G. A. Reed, artist

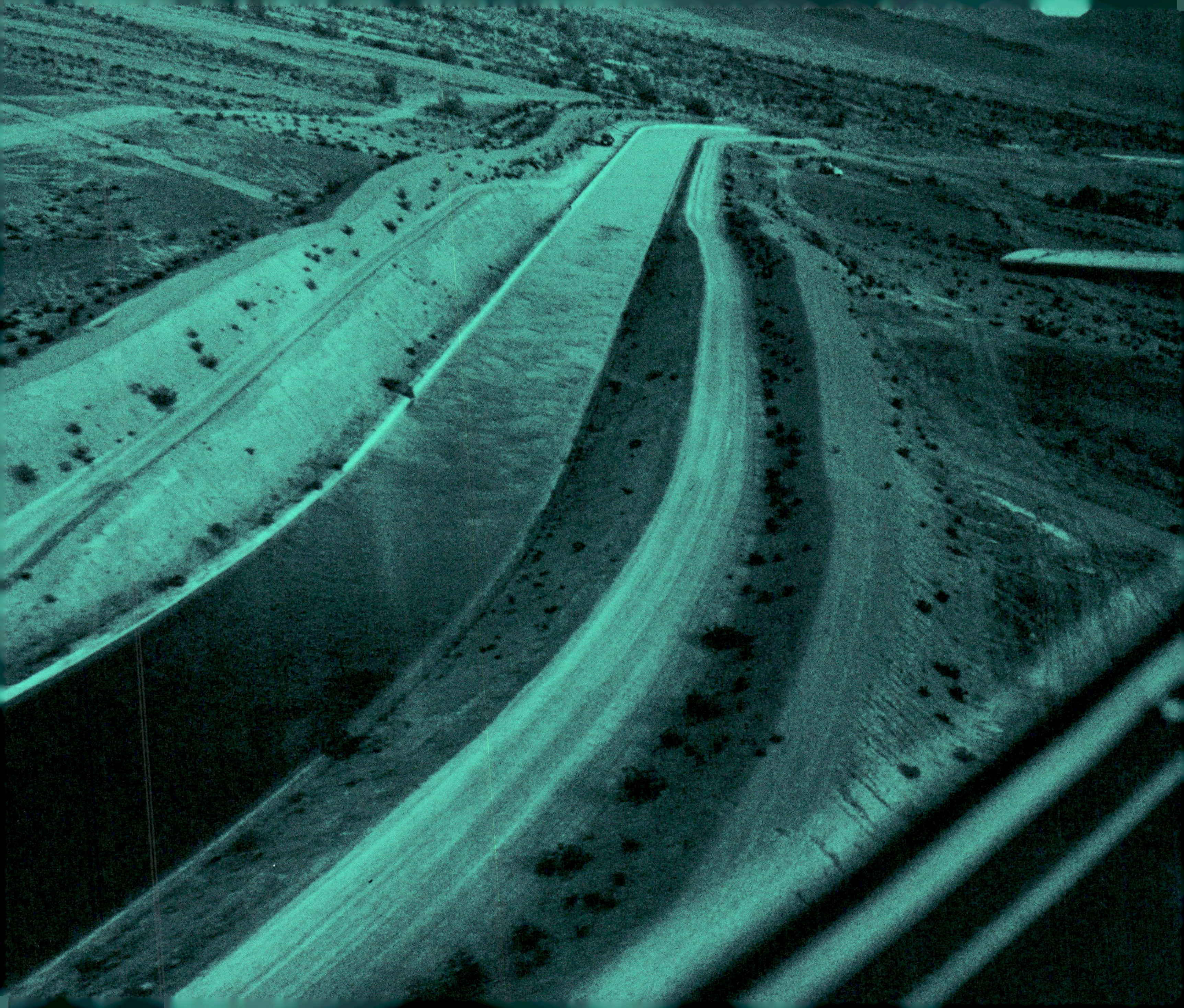

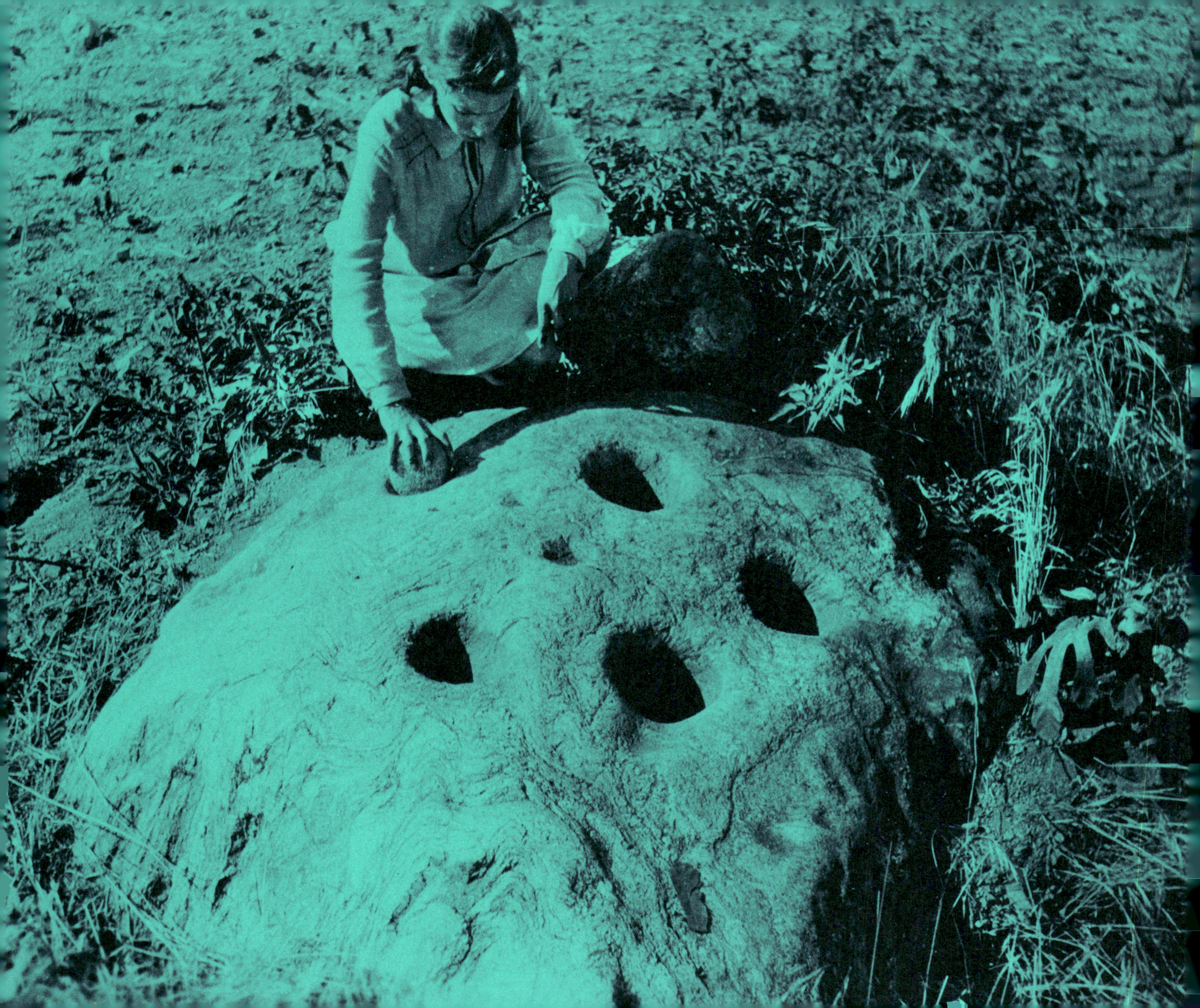

Ground

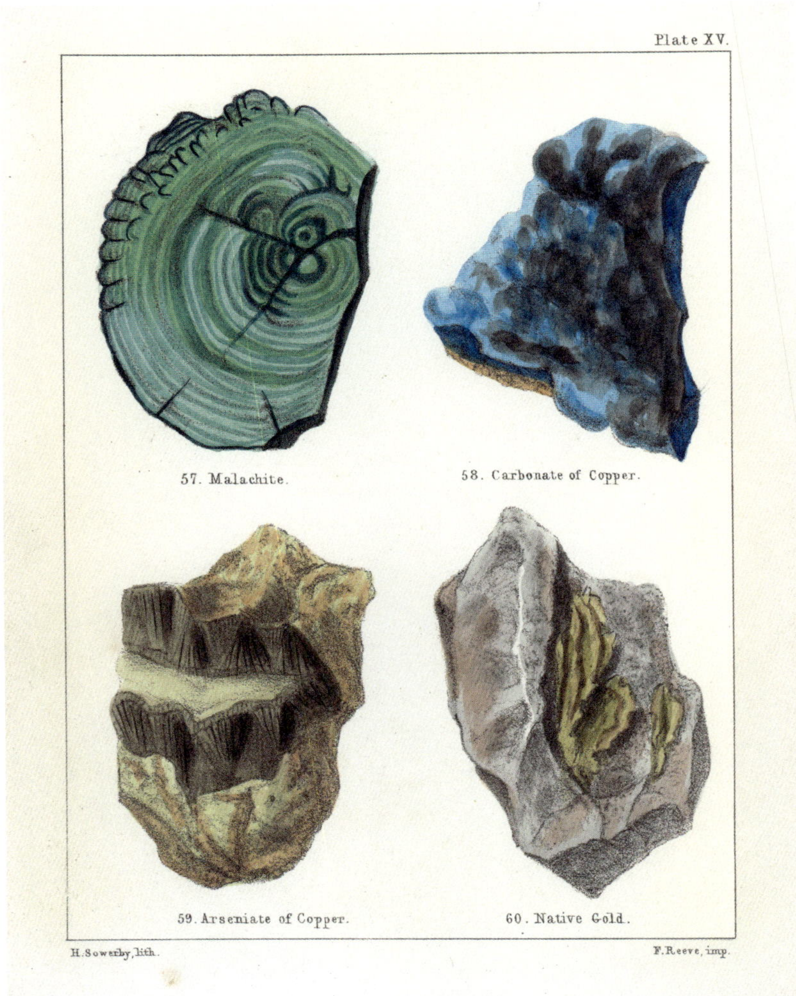
Plate XV.
57. Malachite.
58. Carbonate of Copper.
59. Arseniate of Copper.
60. Native Gold.
H. Sowerby, lith.
F. Reeve, imp.

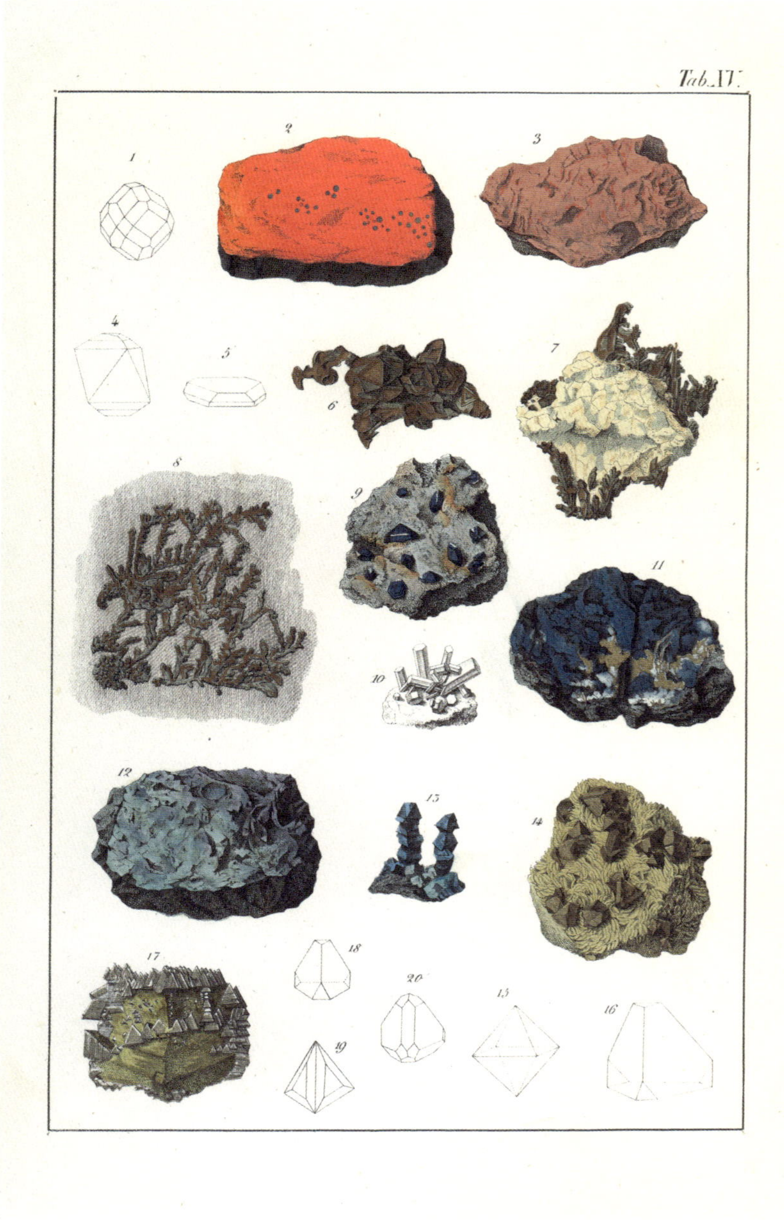
Tab. IV.

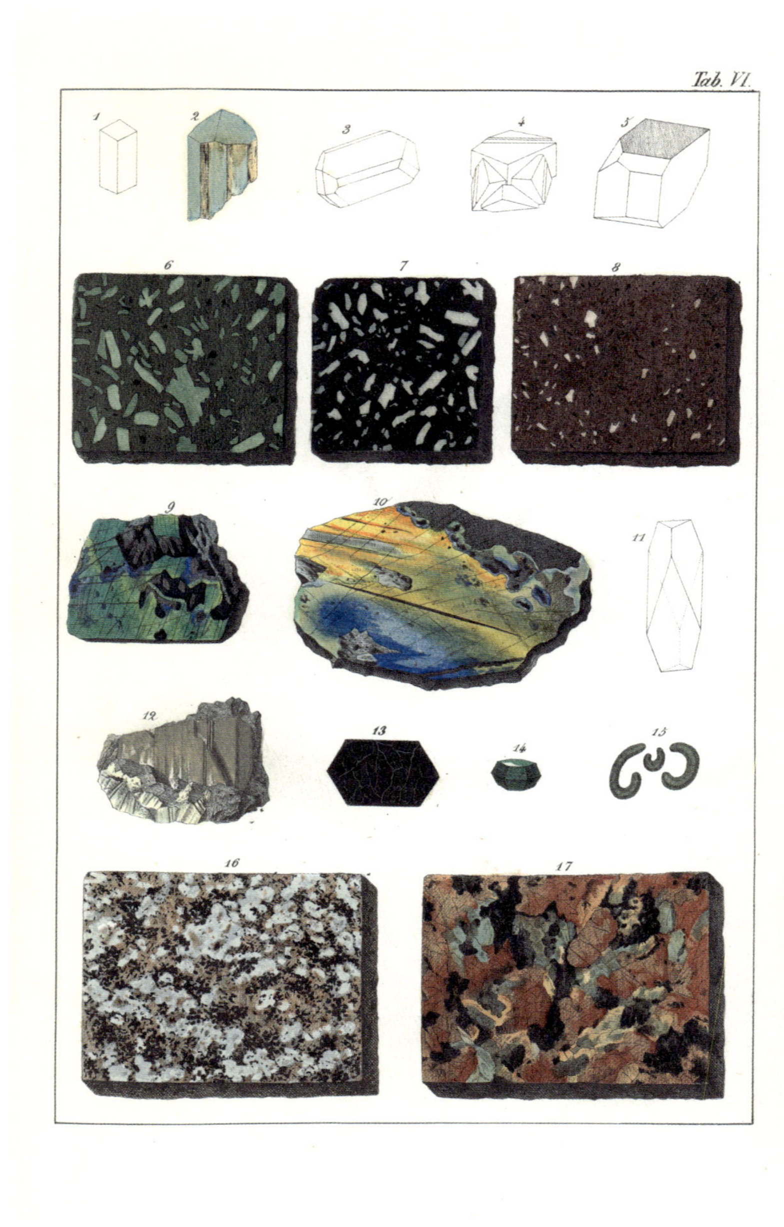
Tab. VI.

↩ *Popular Mineralogy*, 1850. Henry Sowerby (1825–1891)

← *The Mineral Kingdom*, 1869. Johann Gottlob Kurr (1798–1870), author; John van Voorst (1804–1898), translator

The journal of José Longinos Martínez, a surgeon and budding naturalist, recounts his expedition through California in 1792 on behalf of the Spanish crown. He went seeking commercially viable resources, but his notes cover a much broader range of observations on plants, animals, geology, and the customs and technologies of Indigenous communities. He described ancient bitumen deposits full of fossilized animal remains (known today as the La Brea tar pits), and he thought "California" meant "Lime Kiln," due to the abundance of limestone and the oppressive summer heat. The list seen under the heading "*Reyno Mineral*" (Mineral kingdom), begins with silver, hematite, alum, agate, silica, and clay for porcelain.

⊙ 52–53

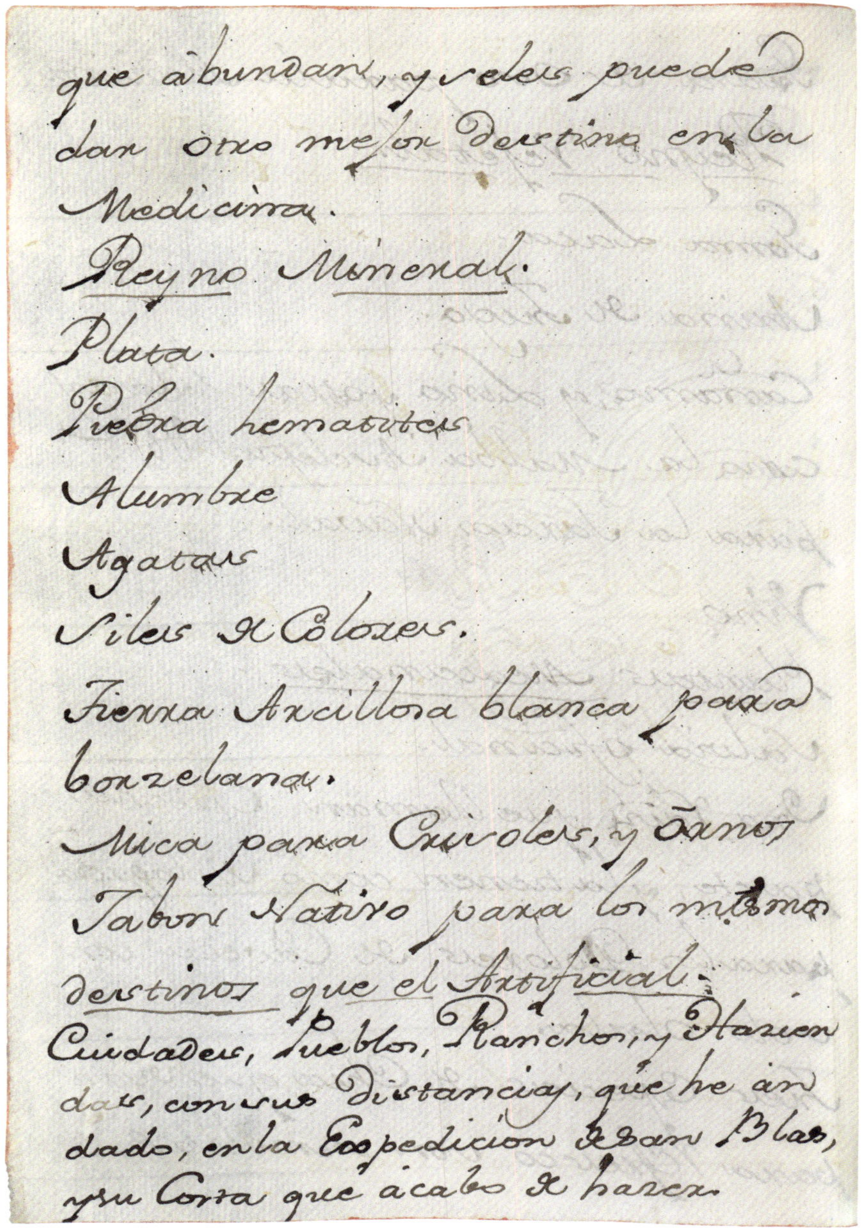
que abundan, y seles puede
dar otro mejor destino en la
Medicina.
Reyno Mineral.
Plata.
Piedra hematites
Alumbre
Agatas
Siles de Colores.
Tierra Arcillosa blanca para
porzelana.
Mica para Crisoles, y paños
Jabon Nativo para los mismos
destinos que el Artificial.
Ciudades, Pueblos, Ranchos, y Hacien
das, con sus distancias, que he an
dado, en la Expedicion de San Blas,
y su Costa que acabo de hazer.

→ Expedition journal, 1791–92. José Longinos Martínez (1756–1802)

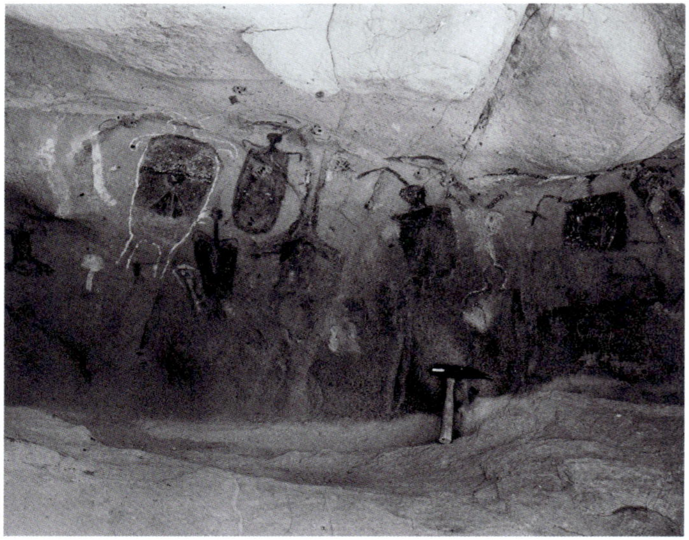

Mineral Pigment

People have painted and adorned everything from cave walls to linen canvases with ochers of all hues: golden saffron yellows, deep burgundy reds, coffee browns, and even warm dark purples. These natural mineral pigments are made from rock composed primarily of iron oxide and clay. To create stable paints, craftspeople carefully select ores, grind and mix them into a smearable paste, and sometimes apply heat to alter the color. This technological and artistic process has been refined for millennia. Archaeologists have identified evidence of ocher pigments in Kenya and South Africa dating back roughly 300,000 years.

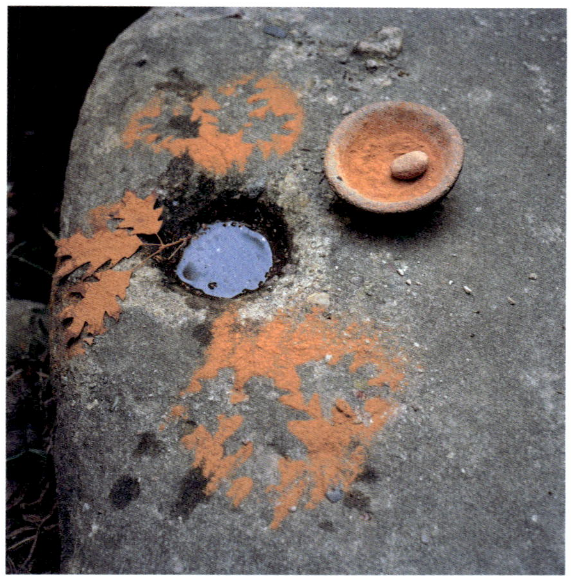

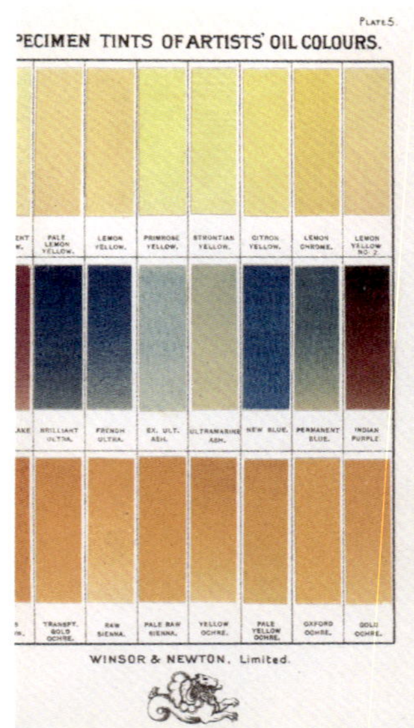

↑ *Pictographs by Chumash or Yokuts people, Kern County, California*, 1904–5. Ralph Arnold (1875–1961)

→ (LEFT) *Well of Moon and Sky—Kotuumot Kehaay*, 2013. Mercedes Dorame (born 1980); (RIGHT) *Specimen tints of artists' oil colours*, ca. 1910s. Winsor & Newton

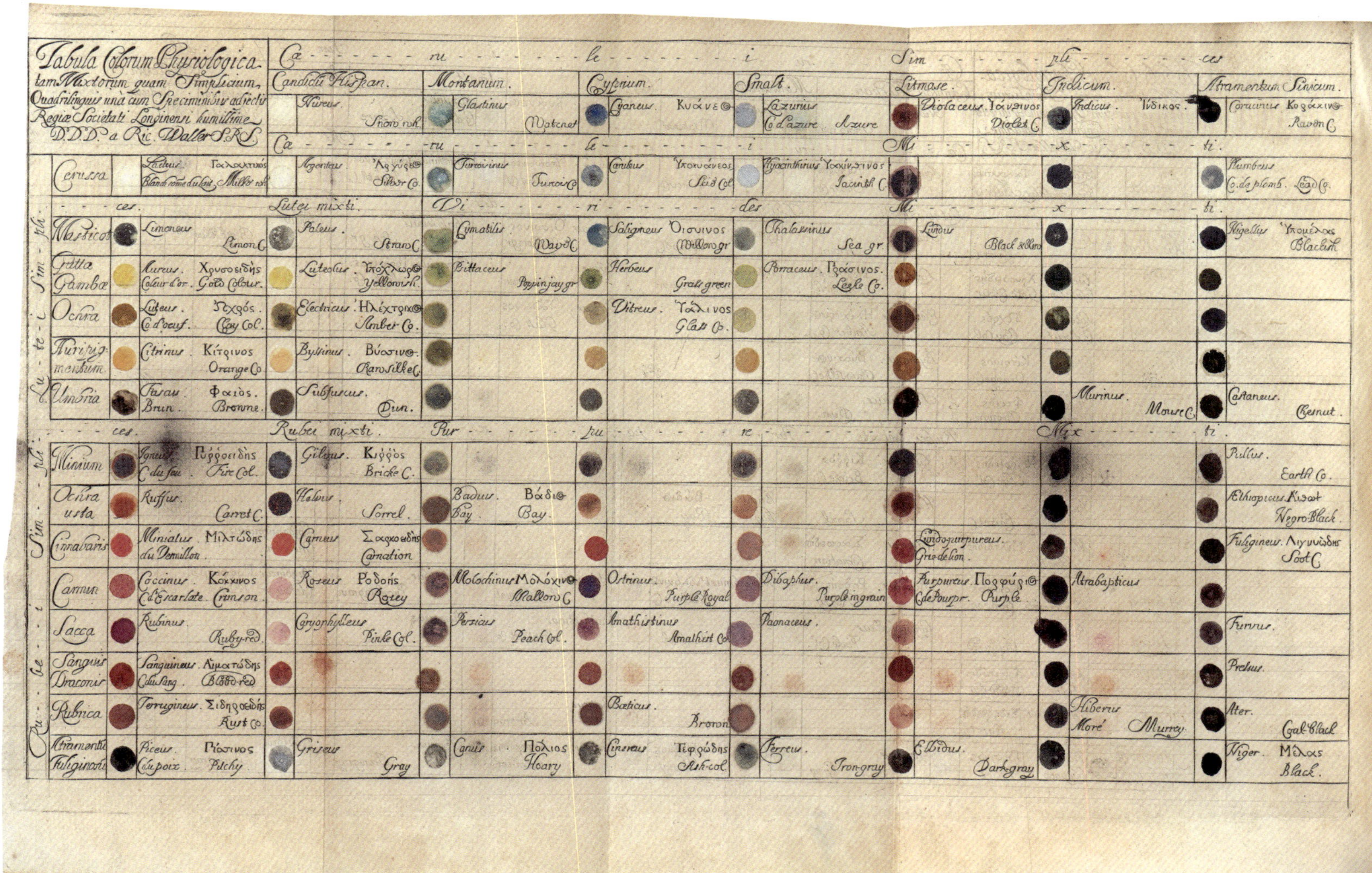

↑ A catalogue of simple and mixt colours in *Philosophical Transactions*, 1686. Richard Waller (ca. 1650–1715)

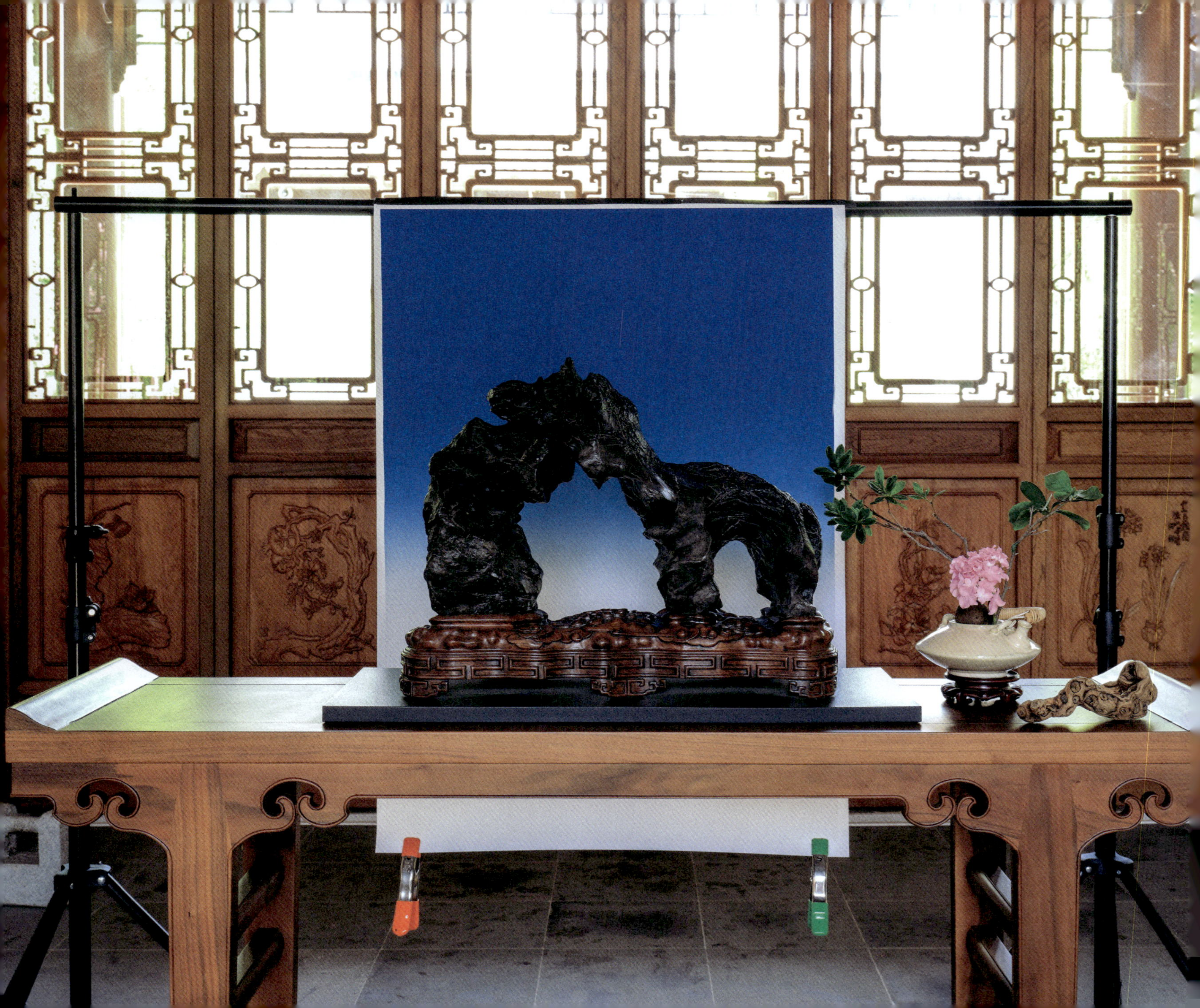

Viewing Stones

Look first. The shape, textures, colors, and sheen of a rock become less static the longer you observe. It teases and scatters light. Its form suggests movement, or maybe a familiar shape, like an archway, a plant, or an animal. Perhaps it evokes a specific place, sound, or memory. For more than 2,000 years in China, scholars and geology enthusiasts have appreciated the subtle power of viewing stones. They encompass the Chinese worldview of essential, animating forces that underpin our environment and material world: *qi* 氣.

← Lingbi stone from Anhui province, China, 20th century

→ *Shizhuzhai shuhua pu* (Ten Bamboo Studio manual of calligraphy and painting), 1633–1703. Hu Zhengyan (ca. 1584–1674)

Stones are kernels of qi and the bones of the earth. In their large state, they are cliffs and crags; in their small state, they are sand and dust.

—Li Shizhen, *Bencao gangmu* 本草綱目 (Compendium of materia medica)

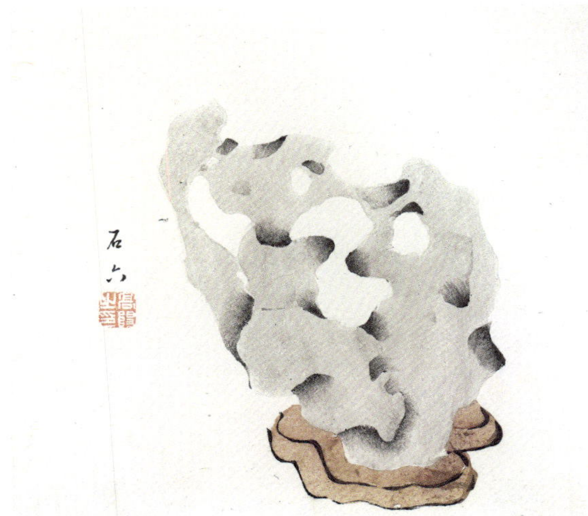

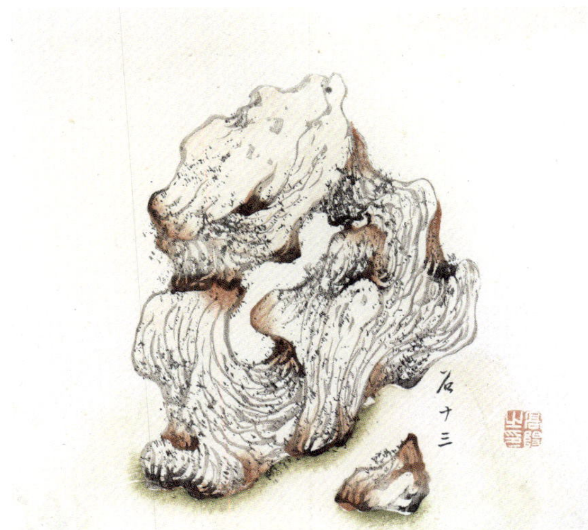

Worlds Unfolding is a slipcase edition containing five books and a booklet.

ISBN (print edition): 978-0-691-28837-6
ISBN (PDF): 978-0-691-28838-3
ISBN (epub): 978-0-691-29521-3
Library of Congress Control Number: 2026930454
British Library Cataloging-in-Publication Data is available.

PUBLISHED BY THE HUNTINGTON
1151 Oxford Road
San Marino, CA 91108
USA
www.huntington.org

IN ASSOCIATION WITH
PRINCETON UNIVERSITY PRESS,
PRINCETON AND OXFORD
41 William Street
Princeton, NJ 08540
USA

99 Banbury Road
Oxford
OX2 6JX
UK
www.press.princeton.edu

GPSR Authorized Representative: Easy Access System Europe - Mustamäe tee 50, 10621 Tallinn, Estonia, gpsr.requests@easproject.com

FOR THE HUNTINGTON
Project management by Jean Patterson
Publication coordination by Mindy Chen
Principal photography by Manuel Flores and Travis Flack
Copyediting by Madeleine Adams
Proofreading by Akiko Yamagata
Art direction and design by Content Object, Kimberly Varella with Gabrielle Pulgar
Color management by Echelon Color, Los Angeles

Printed and bound in Italy by Conti Tipocolor, Florence

The books are typeset in Sharp Serif Text, Sharp Roman, Laica Mono, and Dear Sir Madam and printed on Munken Lynx.

Generous support for *Worlds Unfolding: Science on the Page* is provided by Mario and Therese Molina, Wendy Munger and Leonard L. Gumport, the Rothenberg Family, the Tianqiao and Chrissy Chen Science Initiative, Charles and Valerie Elachi, The Francis Bacon Foundation, the John Brockway Huntington Foundation, and Melvin R. Seiden and Janine Luke.

10 9 8 7 6 5 4 3 2 1

FRONT COVER
Los Angeles Aqueduct, 1966.
Otis R. Marston (1894–1979)

BACK COVER
Telepresence Remotely Operated Vehicle (TROV) *in McMurdo Sound, Antarctica*, 1993.
Dale T. Andersen (born 1955)

FRONT FLAP
Chemical Atlas; or, The Chemistry of Familiar Objects, 1857.
Edward L. Youmans (1821–1887)

BACK FLAP
Mundus subterraneus (Subterranean world), 1669.
Athanasius Kircher (1602–1680)

INSIDE COVER
Kinginzan funai shikioka kasegikatazu (Illustrations of operations inside and outside of gold and silver mines), Sado Island, Japan, after 1788

SECTION DIVIDERS
1 *Second heightening, Aswan Dam, Egypt*, 1931

2 *Earthquake at Santa Barbara steam plant*, 1925. G. Haven Bishop (1879–1972)

7 *The bottom of Cook's Kitchen Engine Shaft* in *'Mongst Mines and Miners*, 1893. J. C. Burrow (1852–1914), author and photographer; William Thomas (1857–1932), author

8 *Diver preparing to guide conduit underwater, Huntington Beach steam station*, 1957.
Joseph Fadler (1924–2013)

29 *View of Stromboli, Italy, from the Sunbeam*, ca. 1874.
Annie Brassey (1839–1887)

30 *Cracks along the highway near Long Beach, California*, 1933.
Larry Lynde

37 *Mundus subterraneus* (Subterranean world), 1669.
Athanasius Kircher (1602–1680)

38 *A. C. Vroman in Petrified Forest, Arizona*, 1895

51 Sketches in notebook of the Geological Survey of the State of New York, ca. 1838. Lardner Vanuxem (1792–1848)

52 *Astronaut and geologist Harrison H. Schmitt working on lunar surface during Apollo 17 mission*, 1972.
Eugene A. Cernan (1934–2017)

65 *Bonners Castle and Ruins of Paradise in Grand Canyon via Ash Fork, Arizona*, ca. 1895. G. L. Rose

66 *General view from the south, Gould & Curry Works, Nevada*, ca. 1865. Lawrence & Houseworth (1851–68)

77 *Live oil seepages near Zumba well, Mene Grande, Zulia, Venezuela*, 1914. John A. Stokes

78 *Surveyors, Los Angeles Aqueduct, Owens Valley, California*, ca. 1906. Frank Rolfe (born 1874)

85 *Los Angeles Aqueduct*, 1966.
Otis R. Marston (1894–1979)

86 *Woman with bedrock mortars*, ca. 1915

95 *John Wesley Powell Expedition near the upper end of the Colorado River*, 1872.
E. O. Beaman (1837–1876)

96 *Upheaval Dome, Utah*, 1961.
Otis R. Marston (1894–1979)

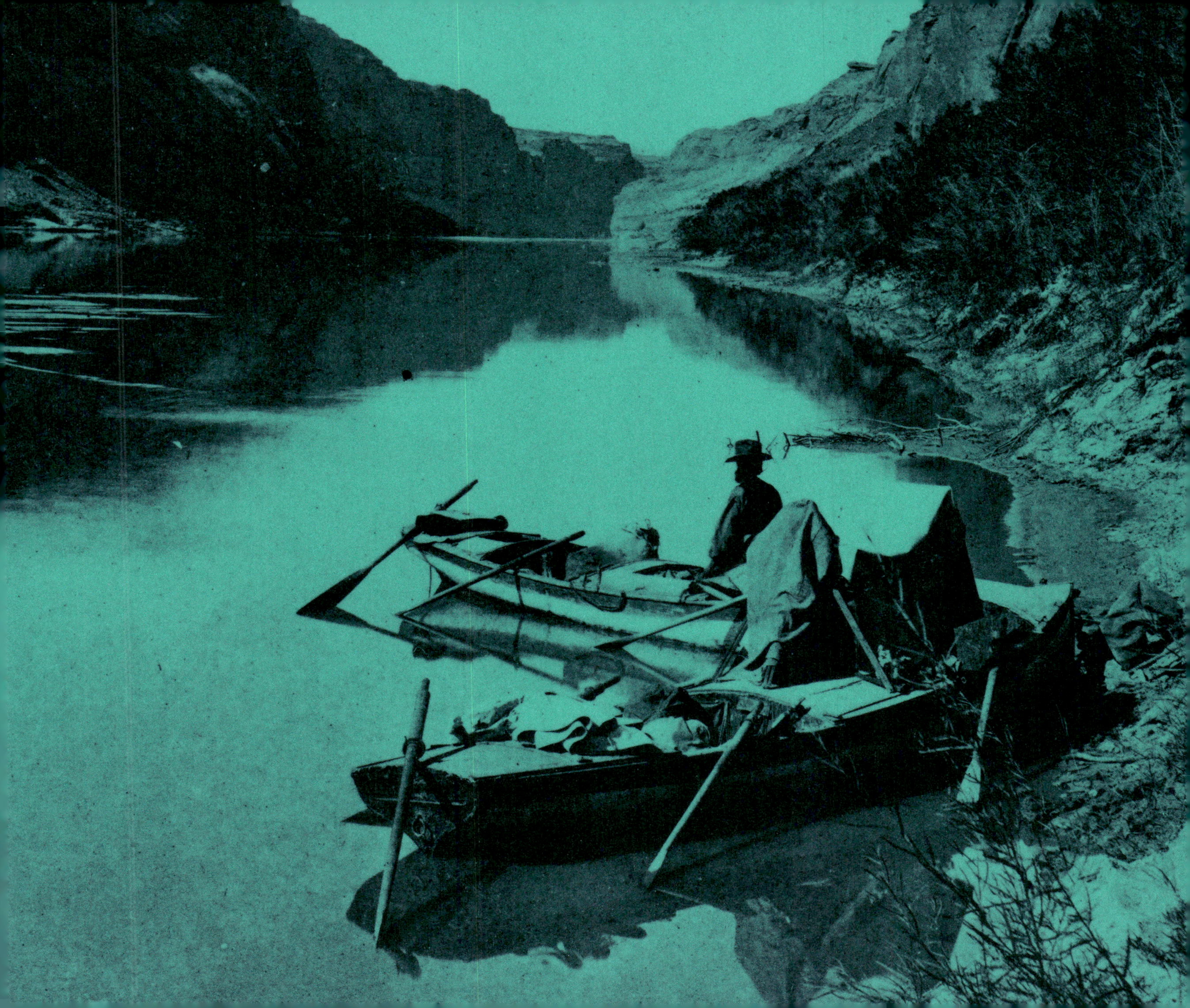

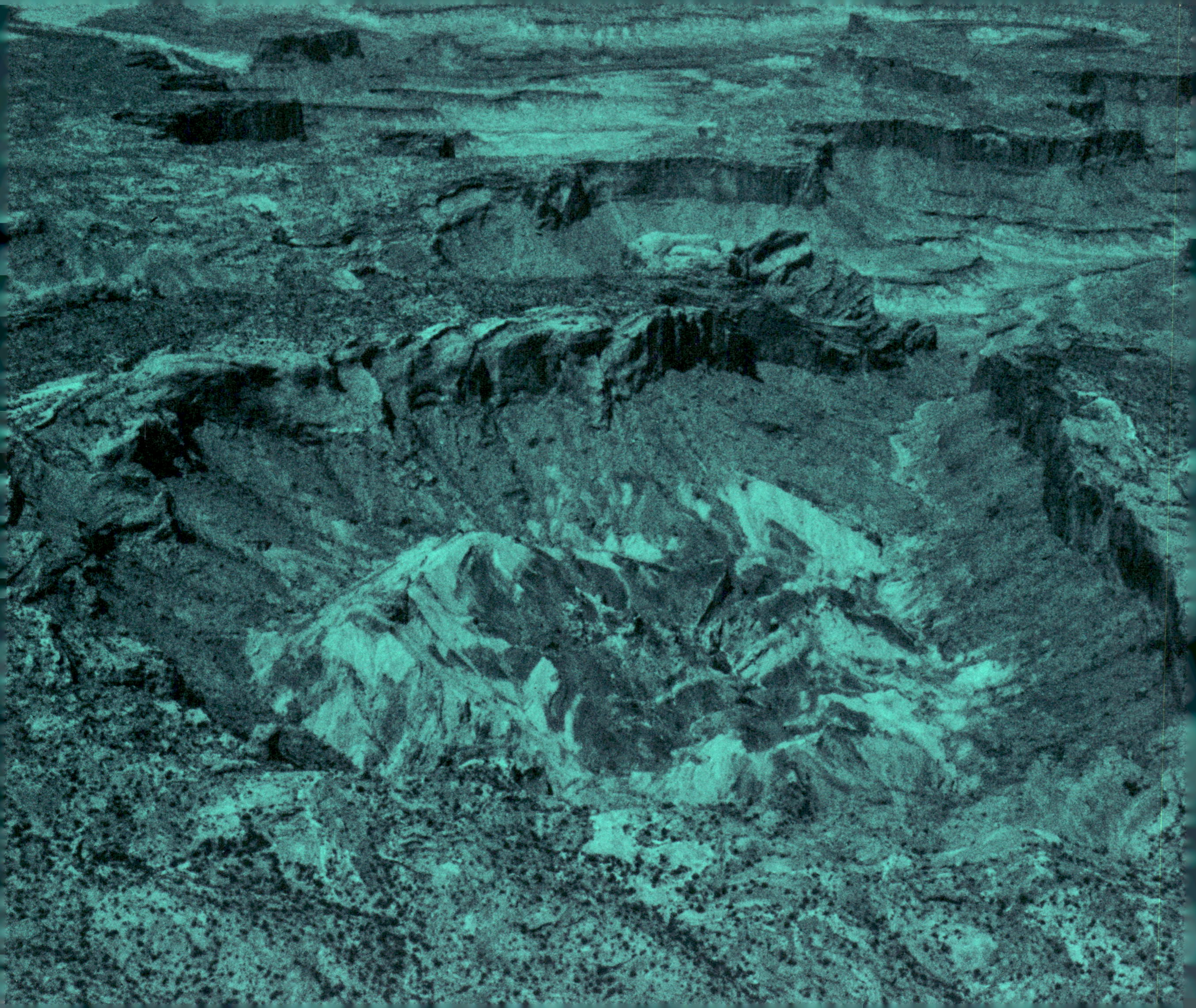